SAGE was founded in 1965 by Sara Miller McCune to support the dissemination of usable knowledge by publishing innovative and high-quality research and teaching content. Today, we publish over 900 journals, including those of more than 400 learned societies, more than 800 new books per year, and a growing range of library products including archives, data, case studies, reports, and video. SAGE remains majority-owned by our founder, and after Sara's lifetime will become owned by a charitable trust that secures our continued independence.

Los Angeles | London | New Delhi | Singapore | Washington DC | Melbourne

Views on *Feminists and Science:*
Critiques and Changing Perspectives in India, Vols 1 and 2

'*If we delve into the philosophy of science, and its methods of inquiry, we recognize the assumptions on which it rests, namely, logical positivism, which believes that science is an objective and value-free enterprise.* [*This volume*] *exposes brilliantly these limitations in the very philosophical basis of science. Gender, caste and class are all implicated in how science is done.*'
–Maithreyi Krishnaraj, Senior Honorary Fellow, Research Centre for Women's Studies, SNDT Women's University

'*One in four scientists in India is female—most hold lower level positions. The obvious question is what's holding women back? Feminists and Science begins the conversation by addressing the epistemologies and practices of science in India. In a series of richly textured essays . . . read together, they are a powerful indictment of the practice of science . . . This book is a must-read for scientists and feminists around the world.*'
–Geraldine Forbes, Distinguished Teaching Professor Emerita, Department of History, State University of New York Oswego

'*Here is a welcome volume, indeed! . . . One of the lessons of the last decades of scholarship is the importance of difference and variation within categories previously thought homogeneous. Gender and science are chief among these. The essays in this volume bring together distinctive perspectives on the sciences among Indian feminists, focusing on sciences relevant to feminists in India, as they are practised in India.*'
–Helen E. Longino, Clarence Irving Lewis Professor of Philosophy, Stanford University

FEMINISTS AND SCIENCE

Thank you for choosing a SAGE product!
If you have any comment, observation or feedback,
I would like to personally hear from you.
Please write to me at **contactceo@sagepub.in**

Vivek Mehra, Managing Director and CEO, SAGE India.

Bulk Sales

SAGE India offers special discounts
for purchase of books in bulk.
We also make available special imprints
and excerpts from our books on demand.

For orders and enquiries, write to us at

Marketing Department
SAGE Publications India Pvt Ltd
B1/I-1, Mohan Cooperative Industrial Area
Mathura Road, Post Bag 7
New Delhi 110044, India

E-mail us at **marketing@sagepub.in**

Get to know more about SAGE

Be invited to SAGE events, get on our mailing list.
Write today to **marketing@sagepub.in**

This book is also available as an e-book.

FEMINISTS AND SCIENCE

Critiques and Changing Perspectives in India

Vol 2

Edited by
Sumi Krishna and Gita Chadha

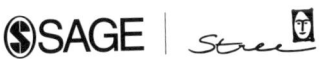

Copyright © Sumi Krishna and Gita Chadha, 2017

All rights reserved. No part of this book may be reproduced or utilized in any form or by any means, electronic or mechanical, including photocopying, recording, or by any information storage or retrieval system, without permission in writing from the publisher.

First published in 2017 by

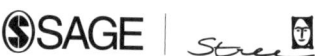

SAGE Publications India Pvt Ltd
B1/I-1 Mohan Cooperative Industrial Area
Mathura Road, New Delhi 110 044, India
www.sagepub.in

STREE
16 Southern Avenue
Kolkata 700026
www.stree-samyabooks.com

SAGE Publications Inc
2455 Teller Road
Thousand Oaks, California 91320, USA

SAGE Publications Ltd
1 Oliver's Yard, 55 City Road
London EC1Y 1SP, United Kingdom

SAGE Publications Asia-Pacific Pte Ltd
3 Church Street
#10-04 Samsung Hub
Singapore 049483

Published by Vivek Mehra for SAGE Publications India Pvt Ltd, typeset in 11/13 pt Baskerville by Zaza Eunice, Hosur, Tamil Nadu, India and printed at Saurabh Printers Pvt Ltd, Greater Noida.

Library of Congress Cataloging-in-Publication Data Available

ISBN: 978-93-81345-19-1 (PB)

SAGE Stree Team: Madhuparna Banerjee, Supriya Das and Neha Sharma

To all those who have inspired feminist and democratic ways of seeing the world. —S.K.

To those marginalized, erased and humiliated by hegemonic truth-making discourse. —G.C.

As this volume was in press, we were saddened to learn of the passing of Dr. Ajita Chakraborty (who figured in Chapter 7 of volume 1) in Kolkata and of Professor Chitra Natarajan (co-author of Chapter 26 of volume 2) in Mumbai. Ajita's pioneering contribution to the then fledgling field of psychiatry and mental health science and Chitra's innovative work in science and technology education will forever be part of the narrative of feminist science in India. –Editors

Contents

Preface   ix

Introduction
    Understanding Gender and Science in India:
    Institutions and Beyond   xi
    Sumi Krishna

    Tracking a Consciousness: Questions, Dilemmas and
    Conundrums of Science Criticism in India   xxxiii
    Gita Chadha

14   *Contextual Empiricism and Local Community: Postcolonial*
      Reflections on Longino   1
      Kanchana Mahadevan

15   *Traditional Knowledge and Feminist Dilemmas: Experience*
      of the Midwives of the Barber Caste in South Tamil Nadu   23
      Meena Gopal

16   *Journeying through an 'Alien Terrain': Feminist Research*
      Methodology and Local Knowledge of Soil Management   46
      Meghana Kelkar

17   *'Blue Flower Mentoring': Interview with Vidita Vaidya*   72
      Gita Chadha, Sumi Krishna and Unnati Tripathi

18   *The Healing Touch: Dr. V. Shanta's Journeys in Cancer*
      Treatment and Care   93
      Kamala Ganesh

19	*Body, Reproduction and Technology: Local Subversions and Global Regressions* Chayanika Shah	121
20	*Feminism and Science: Present-Day Notes for a Feminist Standpoint Epistemology* Asha Achuthan	147
21	*Science and the Making of a New Nationalist Masculinity in Colonial Bengal* Madhumita Mazumdar	175
22	*Fingerprints and Erasures: Mapping the Creative Process in Science* Gita Chadha	207
23	*Science in Architecture and Architecture in Science: A Conversation with Neera Adarkar* Gita Chadha and Unnati Tripathi	232
24	*Gender and Science: The Fiction of Lila Majumdar* Ipshita Chanda	253
25	*Gender, Science and Technology Education in India* Anitha Kurup	278
26	*An Inclusive Science and Technology Education Curriculum at School Level* Sugra Chunawala and Chitra Natarajan	299

About the Editors and Contributors — 324

Table of Contents, Vol 1 — 328

Preface

This book has had a long gestation. In 2006, the Indian Association for Women's Studies (IAWS) began a dialogue on various aspects of feminist knowledge production. Sumi Krishna, feminist science researcher and practitioner, who was then President of the IAWS, initiated a workshop (Bengaluru, 2007) on 'Feminist Perspectives and the Struggle to Transform the Disciplines' to take stock of attempts to critique the approach and methodology of a range of disciplines in India.[1] Many of the presentations were on aspects of gender and science, and so a year later, at the IAWS Silver Jubilee National Conference on 'Feminism, Education and the Transformation of Knowledges: Processes and Institutions' (Lucknow, February 2008), a sub-theme on 'Gender, Science and Technology' was included. The IAWS invited feminist science researcher and lecturer Gita Chadha to coordinate this.[2] The stimulating interactions in 2007 and 2008 helped us identify a group of feminist researchers and practitioners who shared our interest in critiquing science and seeking theoretical and practical alternatives. These researchers and practitioners, with very diverse disciplinary moorings, were invited to submit papers with a view to publication as a book. In 2010, with support from the IAWS this group came together at the Research Centre for Women's Studies (RCWS), SNDT Women's University, Mumbai, for a lively peer-review workshop.[3] We were fortunate also to have among us a doyenne of women's studies in India, Maithreyi Krishnaraj, who had been the first to draw attention to the field of women and science.[4]

As in many feminist endeavours, it is the commitment of the participants that has brought this project to fruition. And our deepest thanks are due to the contributors who wrote, revised and responded to queries and more queries as we worked collectively together. To our own surprise, the collection eventually encompassed 26 essays.

The editors acknowledge with gratitude the IAWS, all the other institutions that have hosted us and our co-coordinators in the workshops at the United Theological College (UTC), Bengaluru (2007), the University of Lucknow (2008) and the RCWS, Mumbai (2010). We also thank all those who contributed papers or participated in these interactions, although not all of them find a place in the published volumes. Suchitra Mathur (IIT Kanpur) and Unnati Tripathi (Mumbai) helped with some of the editing. We owe special thanks to our publisher, Stree, whose belief in this project and keen oversight has been most critical.

It remains for us to thank our families, friends and colleagues for their encouragement and support.

Notes

1 The Bengaluru Workshop, held in February 2007, was coordinated by Sumi Krishna (then President of the IAWS and Southern Regional Coordinator) with M. Indira, Mysore University, Rev. Evangeline Anderson-Rajkumar and Rini Ralte, UTC, Bangalore. (See Krishna 2007. 'Feminist Perspectives and the Struggle to Transform the Disciplines: Report of the IAWS Southern Regional Workshop', Indian Journal of Gender Studies 14, 3: 499–515.)
2 The sub-theme at the Lucknow Conference in February 2008, coordinated by Gita Chadha, then of Russell Square College, Mumbai, and Riddhi Shah, Jawaharlal Nehru University, New Delhi, included papers and a lively panel discussion.
3 The 2010 peer review workshop was held in conjunction with a seminar on Gender and Science organized by IAWS and RCWS at SNDT, Mumbai, coordinated by Sumi Krishna, Veena Poonacha, Gita Chadha and Meena Gopal.
4 See Mathreyi Krishnaraj. 1991. *Women and Science: Selected Essays*. Mumbai: Himalaya Publishing House.

Introduction

Understanding Gender and Science in India: Institutions and Beyond

Sumi Krishna

Science is a part of society and also its product. As Richard Lewontin (1993: 3) said, 'Scientists do not begin life as scientists, after all, but as social beings immersed in a family, a state, a productive structure, and they view nature through a lens that has been molded by their social experience'. How the 'universal' language of science is shaped by class-caste locations and gendered perspectives are tantalizing questions for feminists in India. The science establishment and scholars of science may acknowledge that scientists are social beings, viewing nature through a social lens. Yet, this rarely leads to recognizing the interwoven and embedded inequities of gender, class, caste, religion and language in scientific organizations and in the very process of knowledge creation. Feminists and Science is a political project that brings the discussion from the margins of both science and gender to the centre. In order to understand how the politics of science and gender plays out, feminist critique needs to focus on science as an institutional system of knowledge production and on its role in perpetuating gender inequities (see Harding 1986; 1991).[1] Changing systemic practices is a step towards the ultimate goal of transforming the very nature of scientific disciplines (Krishna 2007).

Such a progressive critique encompasses but goes further than the conventional Women in/and Science approach.[2] Till very recently, gender issues rarely figured in histories of science in India.[3] Nor has science been a major area of concern in the rich feminist and women's studies literature of recent decades.[4] India's

Sixth Five Year Plan (1980–1985) included a section on 'women in science' (GoI 1980) and since then the main focus of policy has been on increasing women's enrolment in science.[5] The Science and Technology Policy (GoI 2003) aimed to empower women in science and technology activities by ensuring their full participation. In practice, however, this has meant schemes for helping women make up for presumed attitudinal and practical deficiencies in themselves rather than recognizing the gendered structure and culture of scientific research institutes in India (see Poonacha 2005). Women scientists and social scientists too have either not 'seen' the gendered fabric of Indian science or have learnt to 'hold their tongues',[6] as the young girl in Suniti Namjoshi's *Feminist Fables* (1981) is advised to do.

> And after the Emperor had appeared naked and no one had disturbed the solemn occasion, one little girl went home in silence, and took off her clothes. Then she said to her mother, 'Look at me please, I am an Emperor.' To which her mother replied, 'Don't be silly, darling. Only little boys grow up to be Emperors. As for little girls, they marry Emperors; and they learn to hold their tongues, particularly on the subject of the Emperor's clothes.'

Feminist researchers, working (often entirely independently) in different locations in India, have only just begun to critique scientific epistemologies and practices, seek theoretical and practical alternatives, and explore innovations in pedagogy and education. Hence, it is not surprising that there is so little literature on gender and science in India, and that this is so fragmentary and undertheorized.

In these two volumes, therefore, our purpose is to work towards a more comprehensive and grounded understanding of gender and science in India. How gender operates in scientific organizations is connected with how gender operates in social institutions. Certainly, matters related to the day-to-day career concerns of women practitioners of science require attention but many women in most professions also face sexism at work and must struggle for equality. How is science different? Our contention here is that we need to understand the linkages between the scientific and the social and examine how science represents gender, how the disciplines and methods of science are shaped by gender, and how the gendered structure and culture of scientific organizations influences practice. We draw upon the international feminist debate on science studies (see also Chadha 1997) and our perspective is grounded in our specific Indian locations. The 26 essays that make up this project are broadly grouped around three major themes (although some of the essays deal with issues that cut across more than one theme). Volume 1 critiques the epistemologies and practices of science; volume 2 explores conceptual and practical alternatives; and both volumes cover the innovative areas of speculative literature and experimental pedagogy.

Women in Science

Data compiled by the University Grants Commission (UGC 2001) show that the enrolment of women in scientific disciplines in India increased in keeping with the substantial increase in girls' access to education in the late twentieth century. Although there is reported to have been some levelling off more recently, comparing the enrolment figures across different disciplines is revealing (Krishna 2010). The UGC data show that women in medicine accounted for 44 per cent of the total, the same percentage as in the 'arts' (i.e., humanities and social sciences together). Women in science made up 40 per cent, which was slightly higher than women in commerce (37 per cent). In the science-based professions, the ratio of women to men was about one in five: engineering 21 per cent; veterinary science 21 per cent; agriculture 17 per cent. This is on par with, say, law where the enrolment of women was 20 per cent. Agriculture lags behind but veterinary science has made significant strides in women's enrolment: Ramdas (Chapter 9) notes that in the early 1980s, in her class in the Veterinary College, in Hissar, Haryana, there were only two women among 80 men.

Analysing the UGC data, the Indian National Science Academy's Mehtab S. Bamji Committee on women in science and the Research Centre for Women's Studies of the SNDT University, Mumbai (INSA 2004; Poonacha and Gopal 2004) have pointed out that in some states—Goa, Kerala, Punjab and Puducherry—more women than men were studying science. Furthermore, contrary to common assumptions, data indicate that there is very little attrition from the undergraduate to the postgraduate levels although there is horizontal segregation by discipline. A widespread observation is that there are indeed more women in the biological sciences and medicine than in the physical sciences and engineering, which is a worldwide pattern (see NSF 2011).[7]

Significant numbers of women are involved in the lower rungs of Indian instrumentation, electronics, information technology and other technical fields. There is little comparative data available, however, on the numbers and working conditions of women technicians, laboratory attendants, paramedics, and so on. And we know little about how class, caste, religion and language are interwoven with gender in determining women's careers in science and science-based professions. In what circumstances do class and caste override gender? The INSA study (2004: 65) notes that the 'majority of respondents were from urban Hindu forward-cast [*sic*] families, from English medium schools'. If the women in science are mainly from privileged upper class and caste backgrounds, how does this compare, say, with men from a less privileged class caste background? We do know that compared to the number of women graduating in various scientific disciplines, there are far fewer women working in science. There are also significantly fewer women in scientific research than in science teaching. There is vertical gender segregation, with many more women at lower levels, and there are also gender issues related to decision making and management of organizations, all of which are largely controlled and led by men.

It is evident that a 'critical mass' of women practitioners in a profession (science or any other) does help change certain conditions of work for women (provision of toilets at the workplace, maternity leave and such like), but this does not necessarily affect the deep-rooted gender discrimination embedded in the system, and may even exasperate it by triggering male backlash (see Shastri, Chapter 6). The government's pro-women initiatives to provide girls with role models to alter perceptions about science or to train women in areas considered 'masculine', or in which women are perceived to have 'special skills', operate in hostile conditions of male domination. Therefore, deeply gendered assumptions are not questioned; gender sensitization/orientation initiatives for men are rare. Institutionalized gender bias in knowledge systems, organizational structures and practices are simply overlooked. Programmes based on the assumption that the problem is personal and individual have a limited impact in bringing about institutional political change. As Poonacha says, 'While overtly claiming to extend access and outreach, the implicit policy motivation is the maintenance of an exclusive social order. These exclusions (particularly of gender) are also likely to be increased in the current economic compulsions of global capitalism' (2005: 246).

In sum, it is not a matter of numbers alone. Proportionately, there are more women at lower levels. Women tend to cluster in certain disciplines and fields that are perceived as being related to women's conventional 'caring roles'. They face obstacles in choosing scientific careers, especially as researchers. They have to overcome familial and social pressures and cope with gender bias in the culture of scientific organizations. Commenting on the women scientists in C. V. Raman's laboratory, Abha Sur (2011: 215–16) wrote: 'The survival of women in the hallowed halls of science has been poignantly difficult, yet their persistence in these halls does not dissuade or dissolve the gender and class inequities embedded in the larger system'. She said feminists have had 'a lingering hope' that the entry of women in sufficiently large numbers into 'traditionally male dominated fields of inquiry would change the institutional biases and, more importantly, the very nature of these fields'. Yet, while some institutional responses to women did change with the gradual increase of women in higher education in India in the late nineteenth century, 'altering the very nature of science would have required a self-conscious affirmation of gender identities by the women scientists in opposition to the coercive womanhood forced upon them by their male colleagues and society as a whole'.

In contrast to the approach of the government and the male leaders of scientific bodies, the key questions are not why there are so few women in science, or why women scientists may not be doing as well as men, or why so few women stay the course and get recognized for their research achievements. The question is how the hitherto 'masculine' culture of science, engineering, agriculture, veterinary science, medicine, affects the intake and performance of women and men; and whether and how the science disciplines and scientific organizations, as they are structured, reinforce gender bias. The gender challenge thus goes far beyond the struggle for access and against overt discrimination.

Critiquing Epistemologies and Practices

Our social and cultural lives are shaped by the intersecting axes of class, caste and gender. Although this is widely recognized, it is often difficult to understand how any given social situation turns around these axes. It is even more difficult to unravel these intermeshed threads within the domain of science, historically and in contemporary practice. Abha Sur (2008; 2011) is among the few researchers to have attempted to do this. The impetus provided by Sur's studies on caste and gender in modern science in India triggered a spontaneous free-flowing e-mail discussion in 2011 among seven feminists, including scientists and social scientists, who grappled with some of these questions (see Chapter 1, Mehta et al.). The layered and nuanced arguments tellingly reflect the dialogical process of knowledge creation, with our understanding evolving from our different locations. Perhaps, as Mary John (see Chapter 1) remarks, we 'know more about gender than we do about caste in contemporary society' but, as she goes on to say, recent work on sex and gender also shows up how much more there is for us to understand—'what indeed does it mean to be a woman or a man?' This is viewed both as a scientific and a social question, and leads to the 'practical' matter of how communications between scientists and social scientists can be enhanced, and whether artists and designers can provide a bridge.

Interrogating the Culture of Science

All of us are rooted in multiple social locations of gender, class, caste and other social categories, and different disciplinary frameworks,

Introduction xvii

which often operate in ways that are not easily discernible, shaping our socio-cultural spaces and insidiously directing how we produce, use and maintain knowledge systems. This makes it difficult to interrogate the culture of science, its epistemologies and practices. The heart of the matter is the construction of 'merit', a politically fraught issue in India. As Mina Swaminathan remarks (see Chapter 1) many leading scientists are not embarrassed by caste prejudice and staunchly believe in a concept of 'merit' that they are unable to define. Most scientists also see themselves as being free from gender bias or treat this as the problem of an individual's 'mindset'.

Jayasree Subramanian's detailed studies (Chapter 2) on the gendered construction of 'merit' in scientific institutions in India reveal the depth of such individual and systemic biases. We need to dismantle the deep structures of science to understand how the production of scientific knowledge may be insidiously gendered. As Sumi Krishna (2009 and Chapter 3) comments, the language and narratives embedded in the cognitive culture of scientific disciplines both reflect and reinforce existing social biases and in turn shape the knowledge, attitudes and practice of the natural sciences in India. Indeed, there seems to be, as Stephen Alter (1999: 1) argues, a 'cross-disciplinary affinity' in the language of science and the culture of an age: 'These seemingly natural metaphors—half-conscious bonds of logic between distinct fields of knowledge—draw upon the aesthetic sensibility of a given time and place'.

For example, the history of psychology in India (a relatively young discipline), resonates with the aspiration to model itself on the positivist methodology of the natural sciences, based on measurements and quantification. Despite the spurt in women-centric studies, psychological research and practice are 'value-laden', uninformed by a feminist perspective and unmindful of the social context and the researcher/practitioner's own location says U. Vindhya (2011 and Chapter 4, see also Davar ed. 2001). Yet, the feminist perspective itself may need to be shaken from its own normative assumptions as Anita Ghai and Rachana Johri (Chapter 5), who are located at the intersection of psychology, disability and gender, seek to do. The differently-abled woman challenges and extends feminist concerns related to the body, unsettling the feminist understanding of women's reproductive choice by exposing the eugenic ideal inherent in contemporary advances in reproductive technology.

Feminist practitioners of science and science-based resource management view the struggle against institutional gender stereotyping and bias as deeply ideological (see Krishna and Kulkarni 2013; Zwarteveen, Ahmed and Gautam 2012; Lahiri-Dutt 2006). Seema Kulkarni's case study (2013 and Chapter 8) of women water professionals analyses how egalitarian practice is undermined by gendered spheres of knowledge, concepts of 'hard' and 'soft' sciences and women's exclusion from techno-centric work. The challenge lies not just in societal culture but in the lack of transparency and accountability within scientific organizations and in government policies that reinforce traditional gendered notions of caring responsibilities. That this is so despite the claim to praxis based on unprejudiced reason makes science and scientists 'more culpable', as astrophysicist Prajval Shastri argues (Chapter 6, see also Jaiswal 1993; Gupta and Sharma 2002).

A thread that runs through the essays is the need for a practice that intermeshes the scientific and the social. Ajita Chakraborty, the first woman psychiatrist to practise in India, who began working in Kolkata in 1960, tells Mandira Sen (Chapter 7) that academic psychiatry is so strongly biological and medical that it 'is almost devoid of social and human dimensions'. For a new orientation to emerge, Chakraborty says, we need to be more aware of our own experience and critique western teaching on mental health. The significance of the local cultural context in feminist science discourse and practice is addressed by several authors (for example, in volume 2, Mahadevan, Chapter 14; Gopal, Chapter 15; Ganesh, Chapter 18; and Achuthan, Chapter 20).

Similarly, Sagari Ramdas (Chapter 9) draws attention to the 'missing millions' of poor women who are traditionally the main carers of livestock but are entirely overlooked by veterinary science and modern animal husbandry. This has deprived women of their livelihoods, knowledge and social position and undermined food sovereignty. The official response to feminist concerns has simply meant including women in training programmes without any deeper understanding of the location of women and men in the production process or of the gendered implications of global resource appropriation.

Theoretical and Practical Alternatives

Many of the conceptual and empirical questions raised in volume 1 are addressed in volume 2. Local knowledges are layered and complex but not always democratic; traditional community institutions may be an alternative to hegemonic forces but may also reinforce patriarchy (see Krishna 2004). For feminists the tension between the local community as both emancipatory and patriarchal is not easy to resolve. This is apparent from Kanchana Mahadevan's discussion of 'community-oriented epistemology' from a feminist postcolonial perspective (Chapter 14). Taking this further Meena Gopal (Chapter 15) unpacks the politics of knowledge-making in postcolonial India and the 'feminist ambivalence' to local caste-based occupations such as that of women healers (dais, mid-wives). The 'alternative' practices of women healers give women greater control over their health and bodies, and create shared spaces that do not have an equivalent in modern allopathic medicine (see also Achuthan, Chapter 20). Resonating with these concerns is agricultural scientist Meghana Kelkar's attempt to 'transgress the boundaries of the natural sciences' (Chapter 16). Tracing the epistemological roots of her research and finding her way into the world of social enquiry is a 'journey through an alien terrain' but 'an essential and conscious exercise'. Yet as Kelkar demonstrates, science can indeed be done differently.

Doing Science Differently?

Most practising feminist scientists like behavioural biologist Vidita Vaidya (Chadha, Krishna and Tripathi, Chapter 17) believe that being feminists affects them as persons but not how they view their data. A scientist's gender may not determine how she interacts with and mentors her team but the practice of science has to be 'culturally sensitive' says Vaidya. This is akin both to Chakraborty's attitude to mental health (Chapter 7) and to pioneering cancer surgeon V. Shanta's 'home-grown' approach at the Cancer Institute, Chennai. Kamala Ganesh's biographical essay on Dr. Shanta (Chapter 18) shows how, in the first decades of Indian independence, behind the technological and organizational strength of the institute, there was 'an ethic of care' of its two founding women: Dr. Muthulakshmi Reddi and Shanta. Yet, this was not simply a traditional 'feminine' ethic but was transmuted in complex ways (see also Anandhi 2008).

Advances in reproductive technology, particularly in relation to the management of women's fertility has for long been a deep concern for feminist activism and research, both internationally and in India. Chayanika Shah (Chapter 19) expresses her disquiet about the feminist demand for 'value-free' fertility technologies, birth control and assisted conception, which are based on the scientific model of 'understanding leading to control and intervention'. Do feminists essentialize the female body and by doing so reinforce binary constructs of gender? If as Shah says, a feminist science can only emerge from feminist conversations, and experience has to be the 'vantage point for knowledge-making' as Asha Achuthan (Chapter 20) argues, then it is lived experience that yields connections (as in the shared space between a dai and a woman in labour). Then, could a feminist obstetrician—even one with the empathetic standpoint of the 'outsider within'—learn to 'share this space', and an alternative process of knowledge-making?

If the underpinnings of scientific knowledge are to be decoded, the focus of enquiry has to turn from organizations to wider cultural and creative spheres of knowledge-making. Madhumita Mazumdar (Chapter 21) shows how ingeniously western science was used to construct a new modern nationalist masculinity in colonial Bengal. The elaboration of such a masculine ideal incorporating swadeshi elements, and the scientists' own implication in this project, buttressed the perception of science as uni-gendered. From the creative cultural politics of history we move to the creative process of the individual scientist. Gita Chadha (Chapter 22) argues that intuition is critical in the creative phase between the genesis of a problem and the publication of results, but that aspects of intuition are obliterated in the public sphere because of the requirements of 'rationality and objectivity'. Yet intuition and 'feminine aspects' reappear albeit differently in the rhetorical construction of the individual scientific genius.

Innovations in Fiction and Pedagogy

It is in the realm of speculative fiction that alternatives have been most vividly articulated.[8] For Indian feminists, a historically significant work is social reformer Rokeya Sakhawat Hossain's fantasy world 'Sultana's Dream' (2005, first publication 1905) where women rule and men are secluded. Even as the novel is futuristic in envisioning an alternative science, it is a product of its times in portraying the *feminine* woman as the pivot of this world. (A hundred years on, even the INSA study [2004: 66] falls back on valorizing supposedly feminine qualities of women, concluding that women's presence in science is advantageous to society because they 'bring in a gentler and more humane perspective to scientific research'.) Discussing Lila Majumdar's Bangla science fiction for children, in volume 2, Ipshita Chanda (Chapter 24) shows how these mid-twentieth century stories meld reason and imagination to envision a harmonious ecology, a world where solar energy is conserved and where men and women, young and old, educated and uneducated are all equal. Scientific knowledge may be fragmented but, with the mediation of human agency, feminist utopia and rationalist science converge in a warm glow. A much starker world emerges from Suchitra Mathur's reading of contemporary Indian feminist science fiction (Chapter 13 in volume 1): Rimi Chatterjee's dystopian vision in *Signal Red* (2005) and Vandana Singh's 'feminist re-visioning of "science" itself' in *Distances* (2008). Both works are centred on the scientific institution as critical to the production of modern scientific knowledge. The power games and reductive outlook in Chatterjee's highpowered state-sponsored institution leave no space for a social understanding of the world. Singh, however, seeks to build an imaginative alternative framework for creating science, for producing knowledge that as Mathur comments 'is not exclusively masculinist', even if this is 'compromised eventually'.

The Pedagogy of Science

Ramdas (Chapter 9) called for veterinarians, especially young veterinarians, with the capacity and sensitivity to analyse and understand 'gender dynamics'. Using a novel pedagogical approach, architect Neera Adarkar (Chadha and Tripathi, Chapter 23 in volume 2) has attempted to make architects aware of such dynamics in the construction of the built environment. In volume 1, Mina Swaminathan (Chapter 10) and T.K. Sundari Ravindran (Chapter 11) also describe innovative pedagogical strategies to enable professional scientists acquire a keener awareness of gender. Swaminathan feels conventionally trained agricultural and biological scientists find it difficult to grasp concepts such as social context, difference and relativism. She sees a deep divide and hierarchy in the system, between disciplines, between theory and practice, between the cognitive, analytic, abstract and quantitative, on the one hand, and the affective, intuitive, practical and qualitative, on the other. Reflexive pedagogic methods partly succeeded in a non-government research institution, but not in the formality of an agricultural university.

Gender is a very important determinant in health, with both sex and gender contributing to large-scale morbidity and mortality, yet as Ravindran points out (Chapter 11), medical education in India has been most resistant to change. She recounts a successful project to enhance gender sensitivity among medical educators who then acted as catalysts in altering the formal undergraduate curriculum. Formal curricular change takes time, but motivated teachers can do much to incorporate gender in their teaching. This is not simply a matter of differentiating data by sex but requires a rethinking of research questions and design. But the more challenging problem, as Ravindran says, is the commoditization of medical education and the privatization of health care. Can gender justice be achieved without social justice?

In comparison to the pedagogical approaches of Swaminathan, Ravindran and Adarkar, which are geared to professional practice, the teaching of Feminist Science Studies (and Chadha, Shah, Chapter 12 in volume 1) to Women's Studies students in India has emerged from the feminist discourse on Science Studies. Like other innovative pedagogical approaches it is interdisciplinary but its interface with Science Studies is reflected in its more theoretical orientation. The unique course has sought to occupy a middle ground between 'anarchism and scientism'.

The importance of an inclusive science and technology education is put forward, in volume 2, both by Sugra Chunawala and Chitra Natarajan (Chapter 26) and by Anitha Kurup (Chapter 25). Kurup argues that diversity is the key to scientific progress and would enhance the quality of learning. Unequal access because of gender, class–caste and rural–urban inequality has to be addressed. Women scientists themselves are a diverse group including those who are engaged in research, those who are pursuing non-research careers, and those who are not employed. Chunawala and Natrajan advocate for technology education to be introduced at school. They argue that technology education—as knowledge, design and 'making'—would prevent the gendering of technology at later stages.

Long and Winding Paths Ahead

We began by seeking to critique epistemologies and practices, explore theoretical and practical alternatives, and highlight innovative ways of doing science differently. Interdisciplinary approaches beckon (see Schiebinger and Schraudner 2011) but we have only just begun clearing the thickets; the paths ahead are long and winding. Core assumptions (like the 'rational man' in economics) have long resisted change and methodological approaches are a major challenge. The outstanding problems, in my view, include the need to encompass plural dialogic modes of knowledge-making, interrogate the thrust towards predictability and quantification, and negotiate the ground between the local and the global.

Methodological Issues

Across disciplines, the 'raw' data being collected has increased phenomenally; computers have made compilation relatively easy and the language of numbers is highly persuasive.[9] Many social phenomena, however, cannot be measured; concepts in geography or ecology cannot always be verified experimentally; and mathematical models may be far removed from the world as it exists. Particularly in the human, social and life science disciplines, rigorous qualitative methods, the accumulation of observations and case studies, the perception of patterns and the complex synthesis of interpretation, analysis and theorizing have contributed to understanding. Yet, the belief that only quantification and predictability provide rigour has a strong hold. A recent meeting of the American Association for the Advancement of Science included a session on using 'predictive tools' from the physical sciences to understand human systems and social phenomena (AAAS 2013). Casual observation in different Indian institutions indicates that the lack of maths skills is being used as a convenient criterion, if not a political tool, to filter students' entry into nonmaths fields of higher education.[10] The social (and gendered) implications of such gatekeeping are yet to be analysed.

In *Signal Red*, Chatterjee's fictional critique, patriarchy is the fulcrum around which the scientific institution is organized, and a reductive, binary worldview is epitomized by scientific methodology. Mathur (Chapter 13) draws parallels between this and Shiva's ecofeminist critique of science (1988; 1989). A male scientist in the imaginary institute asks a woman (Chatterjee 2005: 50–51):

> 'Are you a scientist?'
> 'Yes,' said Anu.
> 'What's your field?'
> 'Sociology.'
> 'Oh!' he chuckled. 'That is arts.'
> 'It's a social science, sir.'
> 'Nonsense!' . . . 'You can't call it a science if there's no predictability.'

This exchange reflects the assumption that science caters to a human need for predictability, and that scientific methodology provides a single uniform template for all kinds of enquiry. Indeed, a dictum attributed to the nineteenth-century physicist William Thomson (Lord Kelvin)—on the facade of the University of Chicago's Social Science Research building—says: 'If you cannot measure, your knowledge is meager and unsatisfactory.' A century after Kelvin, Thomas Kuhn (1961, 1962) pointed out that scientific advances were based on 'paradigm shifts' rather than on improvements in the technology of measurement. It is now well accepted that a paradigm reflects the socio-cultural milieu in which it emerges. It has also been argued that there is a human need for uncertainty, that natural science itself has exposed us to 'ever new uncertainties', and that it 'is a very dubious assumption that predictability is universally good' (Aubert 1965: 118–19). Today, predictability does not mark newer fields of knowledge whose content is increasingly complex.

Within the Indian institutional framework, however, cutting-edge science is an elite research activity with little influence on the majority of conventionally educated practitioners-medical doctors, veterinarians, engineers, agricultural scientists, and so on. It is mainly with these professionals that gender/women's studies researchers interact. This reinforces a view of science that some scientists themselves may no longer hold. So, feminist critiques of positivism and reductive analysis in science also need to engage with the developments taking place within science that go beyond binary interpretations of the world to encompass chance, uncertainty, dynamism, plurality and complexity.

Critical, Reflexive Ethic

Elsewhere, I have argued that both modern science and customary practice are implicated in the politics of inclusion and exclusion, of creating and sustaining elites (Krishna 2004). Both need to be critiqued. A critical, reflexive ethic is central to a feminist, gender studies perspective. Such critique must traverse a difficult terrain, perhaps even an alien terrain, as Kelkar (Chapter 16) remarks, because the attempt is to engage continuously with differences without essentializing the local. This requires a mode of understanding and knowledge-making that is dialogic, adopting the perspective of a researcher who is an 'outsider within', recognizing and blending different ways of knowledge-making, and combining a range of research tools to build what I have called 'genderscapes' (see Krishna 2004; 2009; 2012).

Our worldviews and our research are affected by who we are and where we are, but while we may be the products of particular physical, social and cultural environments we need not be, as Swaminathan (see Chapter 1) says, 'prisoners of our milieus—otherwise there would be no revolutions'. Alternative spaces for new kinds of research are being created outside the formal higher education system. Entering the mainstream requires patience and persistence. Subterranean disciplinary currents are not normally discernible and rendering them visible is subversive politics. As I have written elsewhere (Krishna 2007), 'Transformations that threaten to subvert the existing order cannot come without struggle. Yet, unless such a long-term goal is kept in sight it would be all too easy to lose our way in policy measures and interventions that seem to be in women's favour but are no more than safety valves or smokescreens, changing little, if not actually aggravating women's subordination.' Hence, with Abha Sur (2011: 28) we seek 'a philosophy of science which is alert to its democratic potential, but does not simultaneously bury the historical and continuing role of science in legitimizing racial, gender and class/caste discrimination'.

NOTES

1. In common usage the term 'institution' refers to organizations set up for a purpose, like a government bureaucracy or a scientific institute. Here, by institution we also mean social institutions such as the family, market or school. Institutions may be structured in different ways (like a hierarchy or a network) and have different cultures with overt and tacit norms and rules. The structure and culture of an institution act together to include some people and exclude others, reflecting the underlying power relations in the institutional system.

2. Global feminist approaches to women in/and development/environment/science-technology and related fields are politically diverse: liberal Women in Development (WID) approaches emphasize on

women's knowledge and skills but treat women's labour and time instrumentally; radical Ecofeminisms highlight women's supposedly inherent nurturing and sustaining roles; Gender and Development (GAD) draws upon materialist and social relations perspectives to stress the social construction of gender. In the Indian context, I have suggested a typology of Conventional, Celebratory and Gendered approaches (See Krishna 2012 for a succinct account, as also Krishna 1999; 2004; 2009).

3. For example, see Deepak Kumar (2006) on colonial science or David Arnold (ed. 1988) on imperial medicine.

4. In India, Maithreyi Krishnaraj (1991) has argued that organizations rather than families are responsible for the obstacles to women in science. Yet, none of the significant feminist Women's Studies Readers has included science, perhaps because of the paucity of literature (e.g. Ghadially 2007; John 2008; Ray 2012; Banerjee, Sen and Dhawan 2012). Interest in this field has, however, been generated by Subrahmanyan (1998), Godbole and Ramaswamy (2004) in their collection of inspirational biographies of women scientists and by Neelam Kumar (2009; 2012).

5. The Science and Technology Policy 2003, para 5, says: 'Women constitute almost half the population of the country. They must be provided significantly greater opportunities for higher education and skills that are needed to take up R&D as a career. For this, new procedures, and flexibility in rules and regulations, will be introduced to meet their special needs' (GoI 2003; compare GoI 1958; 1983). See the National Policy for the Empowerment of Women (GoI 2001, para 6.11), and I.A.S. 2010; compare Harding (2001): 'Just add women and stir?'

6. Casual observation indicates that most women scientists are resistant to seeing science as gendered. Like other women professionals they do raise career issues: 'leaky pipeline', 'glass ceiling', recognition and awards, work-family balance, and so on. But women scientists seem to be less critical of science than those who may have studied science and later dropped out of pursuing scientific careers. The Indian Women Scientists Association, formed in Mumbai in 1973, seeks to represent women scientists but its thrust has remained on the practical aspects of improving facilities for women in scientific institutions.

7 A recent gender bench-marking study on inequality in science and technology in seven economies included Brazil, India, South Africa, the Republic of Korea, Indonesia, the USA and the EU ranks India the lowest overall because of socio-economic, education and resource access criteria. India is, however, not ranked the lowest in science and technology and innovation participation (Elsevier 2012).

8 In early seventeenth-century England, Francis Bacon's utopian novel, *New Atlantis* (1627), envisaged a state-sponsored research organization where the Baconian method was employed to understand nature, use knowledge collectively for all humankind, enlarge the boundaries of human endeavour and make all things possible. In the popular narrative of science, this quest was transformed into that of the eccentric solitary scientist in his laboratory exemplified by the nineteenth-century science fiction classic, Mary Shelley's *Frankenstein: The Modern Prometheus*.

9 Feminists used the declining sex ratio revealed by the decadal Census of India as a powerful tool in campaigns against female foeticide, leading to a law against pre-natal sex determination.

10 At the outset of his book *Trust in Numbers*, T. M. Porter (1996: x) says: 'I do not claim that quantification is nothing but a political solution to a political problem. But that is surely one of the things that it is, and our understanding of it is poor indeed if we do not relate it to the forms of community in which it flourishes.' See also Gould 1981 and Crosby 1997.

References

A.A.A.S. 2013. 'Predictability: From Physical to Data Sciences'. Session at the Annual Meeting of the American Association for the Advancement of Science. Boston: Feb. 14–18. http://aaas.confex.com/aaas/2013/webprogram/Session5856.html Last accessed 18 July 2013.

Alter, G. Stephen. 1999. *Darwinism and the Linguistic Image: Language, Race and Natural Theology in the Nineteenth Century*. Baltimore: John Hopkins University Press.

Anandhi, S. 2008. 'The Manifesto and the Modern Self: Reading the Autobiography of Muthulakshmi Reddy'. MIDS Working Paper No. 204. Chennai: Madras Institute of Development Studies.

Arnold, David. 1988. *Imperial Medicine and Indigenous Societies*. Manchester: Manchester University Press.

Aubert, Vilhelm. [1961] 1965. 'Predictability in Life and in Science', *Inquiry: An Interdisciplinary Journal of Philosophy* 4: 1–4; reprinted in *The Hidden Society*, 1965. New Jersey,: Bedminister Press.

Bacon, Francis. [1605; 1627] 1966. *The Advancement of Learning and New Atlantis*. London: Oxford University Press.

Banerjee, Nirmala, Samita Sen and Nandita Dhawan, eds. 2012: *Mapping the Field: Gender Relations in Contemporary India*, vols 1 and 2. Kolkata: Stree.

Chadha, Gita. 1997. 'Science Question in Post-Colonial Feminism'. Discussion. *Economic and Political Weekly* 32, 15.

Chatterjee, Rimi. 2005. *Signal Red*. New Delhi: Penguin Books.

Crosby, Alfred W. 1997. *The Measure of Reality: Quantification and Western Society: 1250–1600*. Cambridge: Cambridge University Press.

Davar, Bhargavi V., ed. 2001. *Mental Health from a Gender Perspective*. New Delhi: SAGE.

Elsevier. 2012. 'National Assessments and Benchmarking of Gender, Science, Technology and Innovation'. Elsevier Publishers at: http://elsevierconnect.com/study-women-encounter-inequality-in-sciencetechnology- fields/#sthash.ayoDPPCD. Last accessed 18 July 2013.

Ghadially, Rehana, ed. 2007. *Urban Women in Contemporary India: A Reader*. New Delhi: SAGE.

Godbole, Rohini, and R. Ramaswamy, eds. 2004. *Lilavati's Daughters: The Women Scientists of India*. New Delhi: Indian National Science Academy.

Gould, Stephen Jay. [1981] rev ed. 1996. *The Mismeasure of Man*. New York: W.W. Norton.

GoI. 2003. Science and Technology Policy 2003. New Delhi: Department of Science and Technology, Government of India. http://www.dst.gov.in/stsysindia/stp2003.htm Last accessed 6 July 2013.

———. 2001. National Policy for the Empowerment of Women. New Delhi: Ministry of Women and Child, Government of India. http://wcd.nic.in/empwomen.htm Last accessed 6 July 2013.

———. 1983. 'Technology Policy Statement', New Delhi: Department of Science and Technology, Government of India. http://www.dst.gov.in/stsysindia/sps1983.htm Last accessed 6 July 2013.

———. 1980. '6th Five Year Plan' (as approved by Cabinet), Planning Commission, Government of India: http://planningcommission.nic.in/plans/planrel/ fiveyr/6th/welcome.html Last accessed 6 July 2013.

GoI. 1958. 'Scientific Policy Resolution', New Delhi: Department of Science and Technology, Government of India. http://www.dst.gov.in/stsysindia/spr1958.html Last accessed 6 July 2013.

Gupta, N., and A. K. Sharma. 2002. 'Women Academic Scientists in India', *Social Studies of Science* 32, 5–6: 901–15.

Harding, Sandra G. 2001. 'Just Add Women and Stir?' In Gender Working Group, U.N. Commission on Science and Technology for Development, ed., *Missing Links: Gender Equity in Science and Technology for Development*: 295–307. Ottawa: IDRC, London: IT Publications and New York: UNIFEM.

———. 1991. *Whose Science, Whose Knowledge? Thinking from Women's Lives*. Ithaca: Cornell University Press.

———. 1986. *The Science Question in Feminism*. Ithaca: Cornell University Press.

Hossain, Rokeya Sakhawat [1905] 2005. *Sultana's Dream and Padmarag: Two Feminist Utopias*. Translated with an Introduction by Barnita Bagchi. New Delhi: Penguin Books.

I.A.S. 2010. 'Evaluating and Enhancing Women's Participation in Scientific and Technological Research: The Indian Initiatives', Report of the National Task Force for Women in Science. Bangalore: Indian Academy of Science. http://www.ias.ac.in/womeninscience/taskforce_report.pdf Last accessed 5 July 2013.

I.N.S.A. 2004. 'Science Career for Indian Women: An Examination of Indian Women's Access to and Retention in Scientific Careers: A Report'. New Delhi: Indian National Science Academy. http://insaindia.org/science.htm Last accessed 5 July 2013.

Jaiswal, Rajendra Prasad. 1993. *Professional Status of Women: A Comparative Study of Women in Science and Technology*. Jaipur: Rawat.

John, Mary, ed. 2008. *Women's Studies in India: A Reader*. New Delhi: Penguin Books.

Krishna, Sumi. 2012. 'Genderscapes: Understanding Why Gender Bias Persists in Natural Resource Management': 265–95. In Nirmala Banerjee et al., eds., *Mapping the Field: Gender Relations in Contemporary India*, vol 2. Kolkata: Stree.

———. 2009. *Genderscapes: Revisioning Natural Resource Management*. New Delhi: Zubaan.

———. 2007. 'Feminist Perspectives and the Struggle to Transform the Disciplines: Report of the IAWS Southern Regional Workshop', *Indian Journal of Gender Studies* 14, 3: 499–515.

Krishna, Sumi. 2004. 'A "Genderscape" of Community Rights in Natural Resource Management: Overview.' In Sumi Krishna, ed., *Livelihood and Gender: Equity in Community Resource Management*. New Delhi: SAGE.

———. 1999. 'Involving Women, Ignoring Gender', Paper presented at 'Gender Dimensions in Biodiversity Management and Food Security,' F.A.O. Technical Consultation. Chennai: M. S. Swaminathan Research Foundation, 2–5 Nov.

Krishna, Sumi, and Seema Kulkarni. 2013. 'Gender and Water: Why We Need New Alternatives to Alternative Discourses'. Presented at 'Water Sector in India: A Critical Engagement'. Felicitation Conference in Honour of Prof. Ramswamy Iyer, New Delhi: Forum for Policy Dialogue on Water Conflicts in India, 25–27 Nov.

Krishnaraj, Maithreyi. 1991. *Women and Science: Selected Essays*. Mumbai: Himalaya Publishing House.

Kuhn, Thomas S. 1962. *The Structure of Scientific Revolutions*. Chicago: University of Chicago Press.

———. 1961. 'The Function of Measurement in Modern Physical Science', *Isis* 52, 2: 161–93.

Kulkarni, Seema. 2013. 'Situational Analysis of Women Water Professionals in South Asia': 201–47. In *Women Water Professionals*, edited by Sumi Krishna and Arpita De. New Delhi: Zubaan.

Kumar, Deepak. 2006. *Science and the Raj: A Study of British India*. New Delhi: Oxford University Press.

Kumar, Neelam, ed. 2012. *Gender and Science: Studies Across Cultures*. New Delhi: Foundation Books.

———. ed. 2009. *Women and Science in India: A Reader*. New Delhi: Oxford University Press.

Lahiri-Dutt, Kuntala. 2006. *Fluid Bonds: Views on Gender and Water*. Kolkata: Stree.

Lewontin, Richard. [1991] 1993. *The Doctrine of DNA: Biology as Ideology*. Harmondsworth: Penguin Books.

Namjoshi, Suniti. [1981] 1993. *Feminist Fables*. Melbourne: Spinifex Press.

N.S.F. 2011. 'Characteristics of Scientists and Engineers in the United States, 2006.' Arlington VA: National Science Foundation, National Center for Science and Engineering Studies. http://www.nsf.gov/statistics/seind12/ Last accessed 29 March 2013.

Poonacha, Veena. 2005. 'Uncovering the Gender Politics of Science Policies and Education', *Economic and Political Weekly* (15 Jan): 241–47.

Poonacha, Veena, and Meena Gopal 2004. 'Women and Science: An Examination of Women's Access to and Retention in Scientific Careers', Mumbai: Research Centre in Women's Studies. S.N.D.T. Women's University.

Porter, Theodore M. 1996. *Trust in Numbers: The Pursuit of Objectivity in Science and Public Life*. Princeton: Princeton University Press.

Schiebinger, Londa, and M. Schraudner, 2011. 'Interdisciplinary Approaches to Achieving Gendered Innovations in Science, Medicine, and Engineering', *Interdisciplinary Science Reviews* 36, 2: 154–67.

Shelley, Mary Wollstonecraft. [1818] 1999. *Frankenstein, or, The Modern Prometheus*. Bantam.

Shiva, Vandana. 1989. *Staying Alive: Women, Ecology and Development*, New Delhi: Kali for Women.

———. 1988. 'Reductionist Science as Epistemological Violence'. In *Science, Hegemony and Violence: A Requiem for Modernity*, edited by Ashis Nandy: 232–56. New Delhi: Oxford University Press.

Singh, Vandana. 2008. *Distances*. Seattle: Aqueduct Press.

Subrahmanyan, Lalita. 1998. *Women Scientists in the Third World: The Indian Experience*. New Delhi: SAGE.

Sur, Abha. 2011. *Dispersed Radiance: Caste, Gender and Modern Science in India*. New Delhi: Navayana.

———. 2008. 'Persistent Patriarchy: Theories of Race and Gender in Science', *Economic and Political Weekly* 43, 43 (25 October).

UGC. 2001. 'Tenth Plan Profile of Higher Education'. New Delhi: University Grants Commission.

Zwarteveen, Margareet, Sara Ahmed and Suman Rimal Gautam, eds. 2012. *Diverting the Flow: Gender Equity and Water in South Asia*. New Delhi: Zubaan.

Tracking a Consciousness: Questions, Dilemmas and Conundrums of Science Criticism in India

Gita Chadha

I

This volume is a labour of love. Of each contributor's struggle as a woman, with questions about the nature and meaning of truth—and reality—in their contemporary postcolonial life-worlds. Some are raging, others are quietly persuasive. Some are scholarly, others colloquial. Since the field of feminist science studies is only developing in the Indian context, the volumes are in the nature of initial interventions.

The articles in the volumes traverse over legacies of deeply embedded hegemonic discourses and practices of western modernity in general. The paradigm of modern western science, in particular, defined and determined contemporary notions of what is true and what is real, as well as what is desirable. In our articulations, we challenge and critique given constructions of knowledge and gender that restrict women's ontological and epistemological pursuits of truth-making, in and outside of modern western science.

Further, the articles reclaim women's right to acquire and make knowledge, they assert women's right to be recognized as 'knowers'. The volumes, as a whole, strive to assert an epistemic validity of women's non-hegemonic ways of knowing complex life-worlds. In this sense, the volumes are both critique and reconstruction of the relationship between modern western science as a knowledge-making system and women's claims to science and knowledge. Of course, the uniqueness of the volumes is their situatedness in the Indian context.

Located within the broad field of Feminist Science Studies (FSS), which has been built upon the developments in and of two other fields—science studies and women's studies—the volumes, like the field of FSS itself, inherit and reflect the dilemmas of these areas of study. Interdisciplinary and firmly rooted in people's movements, science studies and women's studies aim at developing a praxis, attempting to transform the world in more egalitarian and sustainable directions. In the Indian context, some of these dilemmas take on a felt particularity which each of us grapples with, academically and otherwise.

Feeling the Volumes

What constitutes knowledge? What is acceptable as valid knowledge? Who is accorded the status of a 'knower'? Why are women and their knowledges subsumed and erased? What is the status of truth in general and scientific truth in particular? How do we, as feminists, navigate our way in and out of a much needed but problematic epistemological and ontological relativism which argues that what is held as true and real is a product of the context in which it is produced? How much of social constructionism do we need to counter absolutist and universalist ideas of the nature of things and of knowledge of these things?

If we abandon science as a tool for critical transformations of patriarchies and other oppressive social structures, what are our alternatives? Is there a humanist temper that we can articulate to substitute or complement the scientific temper?

As postcolonial feminists, how do we relate to the 'indigenous' or traditional systems/ways of knowledge and practice held by women? Given how deeply enmeshed these are in caste and other traditional hierarchies and patriarchies, can we critically reclaim and reinvent these? Can these provide alternatives to scientific knowledge? Can these be integrated with scientific practices? What happens when the state institutions appropriate these?

How do we address questions of incommensurability of theories and practices? How do we handle opposing truth claims? Is it sufficient to talk of epistemic egalitarianism through the diversity discourse? Or do we need to develop a more robust argument for pluralism? Moreover, as feminists from India, must we not keep our attention focussed on 'real world' issues facing women? Is not feminist science criticism an esoteric and elite area of women's studies? Is not technology the main culprit, why critique science?

What do we do with women in science who refuse to recognize gender or feminism? By advocating for more women in science, will we chart the way for a different science? Is science to be held accountable for the number of women in science or is society responsible? Do men and women scientists do different sciences?

How do we defend ourselves from being dubbed by left wing and liberal discourses as being anti-science, anti-development and anti-progress? How do we guard our critiques from being appropriated or manipulated by right wing discourses that critique the enlightenment modernity in order to advance an identity politics dangerous to the fibre of our democracy and our secularism?

II

These and many other questions have engaged us in our journeys and have informed our articulations. As in all feminist engagements, we find ourselves taking different positions on issues in feminism and science too, which depend on our cultural locations and our theoretical preferences. This diversity and pluralism, we celebrate. But what unites us is the desire to develop a culture of science criticism in India, a commitment to sharpen our critical tools in shaping the contours of feminist relationships to science within our local and particular contexts. For this, we have to trace some basic and necessary moves made in science studies in India.

The First Move

Very often academic critiques of natural science begin as methodological concerns from our disciplinary locations. Indian students of the social sciences, particularly of sociology, like all other students of the social sciences, learn that the foundations of the social sciences lie in the same philosophy of the Enlightenment that led to the development of natural sciences. We learn to uphold the value and use of reason for a 'progressive' social transformation. But as Indians, we internalize the more latent and lethal legacy of

Enlightenment philosophies to develop the 'scientific temper' to contest our 'barbaric' traditions, our shabby selves, our murky past. Deeply determined by Nehruvian rationalism and liberalism, the Indian education system teaches us to dream the idea of the Indian nation in scientific terms. We are taught that the success of the method of natural science in explaining, controlling and harnessing nature for the benefit of humankind must inspire all of us to adopt a positivist mindset, which believes in using the scientific method and its best friend— the idea of a scientific temper—to understand and engineer the social world around us. Further, as social scientists, we continue to learn that any credible 'explanation' of society is and must be 'scientific' and that to be scientific we must be 'objective', that is, put a distance between ourselves as knowers and what we wish to know: the social structure, the social processes and the people who make these. And what will help us set this distance between us as social scientists and the world we wish to explain and even transform? It is, once again, the 'method of science'. We learn to imbibe that people's commonsense understanding of the world, the mystic's worldview, the artist's expressions, are differentiated from and somehow 'inferior' to the scientific understanding of the social world, handed down to us by the 'expert' scientist. So also are the explanations of the physical world that came from natural philosophy or other ways of knowing have been proven to be 'inferior' to explanations and interventions of modern science and technology. We learn that the more reductionist we become in our analysis, the closer we get to the elemental truth of the universe—social and natural. We learn this despite several strands in what are called the 'interpretivist' traditions in the social sciences which argue that the method of science is insufficient for the understanding of the social world. The dominant paradigm in the social sciences, we inherit, is a scientistic one that privileges science and scientific method as knowledge-making systems.

Today, as our social world is being driven towards unprecedented changes caused by globalization and a deepening postmodern condition of fragmented selves and societies, the relentless and continuous transplantation of western modernity the growth of free market forces, increased environmental risk, emergence of religious fundamentalism, weakening of liberal democracy and civic society, expanding urbanism and, of course, growing as a nation, we are left wondering what happened to the promise of science, scientific method and scientific temper? We are pushed to see the fault lines in these ideas and take ahead the task of science criticism India. This must be the first move in building a robust relationship between feminism and science in India—pushing the social sciences to evaluate the scientific paradigm for truth, development and progress critically. Through the articles in these volumes, we, as activists and academics, try to address people's movements against the onslaught of 'big science' and our own disciplinary requirements for a critique of science.

The Second Move

If we look back, we reckon that science criticism has been a difficult terrain for social scientists to walk into because the birth of social sciences is deeply tied with the birth of natural sciences. Hence, it became impossible for the social sciences to examine science critically, even when their disciplines required them to do so. Any critical scrutiny of science was foreclosed. For example, in the sociology of knowledge, we as sociologists are asked to scrutinize every way of believing and look for social determinants of all knowledge claims. Be it political ideology, religious belief systems or commonsense worldviews, we are expected to explain these in terms of social factors. But science? Interestingly we are told that science is a 'special case' because its method is untainted by the social and the cultural. And therefore, the 'truth' it produces is universal and beyond all things social, historical and cultural. It is therefore not possible to look for social determinants of science. Inheriting this framework, the sociological mind exempted science from scrutiny. Until the 1960s, the social sciences continued to give a clean chit to science. It is only due to the pacifist movements that implicated science in the making of war and human disasters that science came under the scanner. Gradually, the role of the scientific community in establishing paradigms of scientific knowledge came to be established, thereby challenging the standard notion that scientific knowledge grew out of an internal logic of evidence and theory. The meta-narrative and image of science as contributing to the 'progress of mankind' and the asocial, ahistorical nature of science both stood challenged. It is in the 1960s, perhaps a century and a half after the birth of modern science that sociologists finally began to ask real sociological questions about science.

Today in the sociology of knowledge we have learnt to have the confidence and the 'nerve', to tell ourselves that whatever maybe our preference and 'taste' as modern persons, we as sociologists have to be more rigorous about our understandings of all knowledge systems, putting science at par with other systems like religion, political ideology, art and commonsense. At least methodologically, we have to do so in order to contest the hegemony of science not only upon social science but upon the knowledge base of our worlds. Placing science within the sociological context will help us to see that societal issues about who gets to do science or who gets to gain from technology are deeply linked to the issues of whether science is given special privileges as a knowledge-making system or not. This is not to say that science is like all other ways of knowing but to suggest that we begin to place it on a continuum of ways of knowing. On this continuum, we can strive towards a new dream, a new vision of placing women's knowledges and practices on par and as 'respectable' alternatives to modern western science. Women's status as 'knowers' deserves epistemic credibility. And that is what the articles in these volumes aim to do—rigorously place science within a social context and give credibility to alternative and complementary modes of knowing the world.

By looking for social determinants of science and scientific knowledge as we would for religion or politics or commonsense and by placing it on a continuum with other ways of knowing and doing, are we saying that all scientific knowledge is socially constructed and that there is no difference between science and non-science? Some postmodernists do see science as totally constructed by the social and linguistic machinery. The 'real world'—of nature and body—becomes plastic and totally fluid. While this position has played a tremendous historical role and continues to do so as a provocative and polemical strategy to counter scientific hegemony over our understanding of reality, it is not a sufficient ally for feminist politics. We, as feminists, recognize the need to straddle the two poles of relativism and absolutism in order to churn out mid-terrains. We realize that the ontological world of nature and body constrains our epistemological claims about it reasonably. The volumes seek to highlight two things. One, the ontological world is quite diverse and therefore many pictures of it must be allowed; and two, our epistemological machineries do shape our perceptions, we must be reflexive about these.

The Third Move

The last caveat. In the late 1990s when some of us wrote around the science wars, like many others, we argued that we stand to gain if we develop a critical temper towards modern western, and yes, masculinist science—an argument that had already been made in the 1980s by academics and activists but had not quite caught the imagination of many academics. Further, we said, in order to develop a richer knowledge base for societies, we need to give an epistemic status to people's ways of knowing; however, they may be called indigenous, traditional or local. This position was attacked as being dangerous, especially since it was perceived as playing into the hands of the right wing traditionalists. All attempts to argue that science criticism is aligned more with progressive needs and values both in the academia and the world of realpolitik fell on deaf ears. The task of building science criticism was more rocky than we had imagined, especially in the Indian context when any critique of science was dubbed as anti-science, anti-people, anti-progress, anti-development by both the liberals and the leftists. As we recognized the problems of this polemics, it became clear that arguing for a critique of science and aiming to make space for non-scientific, often traditional and indigenous, ways of knowing in the hope that we have a richer knowledge base was going to require some complicated theoretical moves problematizing the question: what do we do with traditional knowledge-making systems steeped in caste, class and gender hierarchies? Is there a core we need to draw out? Is there a need to 'revive' anything?

We found ourselves uncomfortable with the revivalist discourse that essentializes traditions and also does not work well as strategy. In the search for suitable concepts, we find the concept of hybridity, a concept indebted to postcolonial scholarship, extremely useful. If we do begin to evolve the conceptual framework available in the notion of hybridity of ways and forms of knowing, could we not throw up a conceptual framework more suited to praxis of equality? For example, would it not allow us to access freely new technologies however embedded they are in structures and relations of organization and power that we wish to relinquish along with robustly critiquing the same simultaneously, keeping alive alternative and different ways of knowing things. This polarity on the twin axes of science and anti-science on the one hand and traditional systems of knowledge and scientific ways of knowing on the other has set the dilemmas for feminist science studies too. The articles in these volumes systematically address these.

III

The development of FSS required the maturation of feminist theory just as it required some moves in science studies. It was firstly essential for feminists to recognize that feminisms are about activism and about developing transformative discourses in nonactivist spaces. Such feminisms about activism address very direct issues of women's oppression that deal with the state and civic society, as well as another kind of activism—an activism in theory, in scholarship, in the academia. We realized that the tacit role of an oppressive gender ideology, its symbolic avatar, so to speak, pervades all through the fabric and network of our lives and minds. In the case of science, it became inevitable to see the link between the question of why so few women are allowed in science with questions about the foundational assumptions made in and by science, both about women and about science. It became necessary to link the oppressive objectification of women's bodies through reproductive technologies to the basic relationship between science and technology.

When we began using the category of gender to examine not only how it constitutes the making of individual men and women, how it positions us in relation to each other in patriarchies but also to examine critically its tacit and silent role in shaping our cognitive and imaginative spaces, we sowed the seeds for feminist critiques of science. We began asking questions about the gendering of scientific cosmologies, of scientific worldviews. We began looking deeply into epistemological notions from standpoints of feminisms.

Further, as we destabilized the notion of 'woman' itself, we began seeing how science constructs and legitimizes constructions of the sex, gender and sexuality discourse to reproduce existing hierarchies. It became evident that if we have to address critically issues of how scientific technologies impact on women's lives and bodies, we have to look at the epistemological foundations, methodological assumptions and ontological truth claims of science. In India, this takes on particularly sharp meanings since feminists in India are asked to deliver and address the 'real' problems of women, often by feminists themselves. And that is perhaps why the scholarship in FSS in India has been slow and always linked to issues in development.

As everywhere else, Indian feminists too begin their critique by raising questions about the low representation of women in science. Countering biological explanations and looking for social factors have been common contours that may be developed. Early works on women in science articulated and demonstrated how science, like any other profession, assigns the dual role for women in science, how they find it difficult to climb up the professional ladder due to family responsibilities that come in the way of their professional path. But more radically, feminists are beginning to ask questions about what in science and the scientific community leads to less participation of women in science. The 'openness' and 'merit-based' character of science has been questioned, empirically and logically to show how the masculinized space and practices of and in science exclude women from entering science.

Further, it has become crucial to visibilize individual and groups of women in science, document their lives and contributions, hear their narratives and read them from feminist locations. Articles in these volumes do that. While critiquing the foundational domain of modern western masculinist science, it has become inevitable for us to demonstrate, more clearly, the prefix 'western' to our transnational feminist communities and to underscore the masculinist prefix to male science studies compatriots in India. Articles in the volumes aim to do that by locating science within the colonial encounter and seeing gendered ideologies within these. As we traverse further into the terrain, it has been necessary to look at what are held as women's practices and knowledges and dig out their erasures and appropriations be these in actual systems of practice of childbirth or in the construction of epistemic categories like intuition and reason. We have had to do this from all disciplinary locations—philosophy, history, sociology, psychology, literature and even the natural sciences. And that is where I must close. The difficult dialogue between women practitioners of science and feminists doing science studies has only just begun in order that a measure of reflexivity seeps into the practice of science, from a feminist consciousness. Also as we make and share these articulations through these volumes, we recognize the deep hierarchical systems that tie up systems of stratification with systems of knowledges in our cultures, hybridized and fractured, seeking to hear the echoes from the past and grapple with pressures of the present. Within this context, these volumes document the efforts to critique, reclaim and reconstruct our encounter with modern science.

IV

Departing from convention, I would like to offer as references—more in the nature of an intellectual debt—the names of scholars who have shaped the field of science criticism. Their works have shaped the consciousness of this essay, and some of the essays in this volume. Any young scholar who wishes to enter this field must discover and engage with these works, critically and with vigour.

Find for yourself works of J.P.S. Uberoi, Ashis Nandy and Shiv Vishvanathan, who mounted the most brilliant critique of science from India but missed out on feminism and gender. Discover Vandana Shiva for bringing that in. Whatever the later debates around Shiva's work, it has historical significance for the field of feminist science criticism in India.

Evelyn Fox Keller, Sandra Harding and Donna Haraway—every word of their works has been inspiring in developing robust and radical mid-terrains over critique, reclamation and reconstruction of science from feminist perspectives. These must be discovered. And of course critiqued from postcolonial locations. From a disciplinary location, the works of Thomas Kuhn and David Bloor have opened sociology to science criticism. The sheer vastness, depth and stylishness of their writings is highly endearing.

And then there is the radical not so well-respected maverick philosopher, Paul Feyrebend. One must find him, simply for the joy of polemics. One must also of course move on from the polemics!

And please find–in this volume and elsewhere–the works of Chayanika Shah, Asha Achuthan and Kanchana Mahadevan, my peers, who have also done some great hand holding through their own work.

Works of the above scholars are cited—though not actively, as must be done in the recommended style for academic writing—in every line of the above piece and stand plagiarized! They must be sought and read. I owe all my articulations to them.

14

Contextual Empiricism and Local Community: Postcolonial Reflections on Longino

Kanchana Mahadevan

In certain theoretical contexts, the only reasons for preferring a traditional or an alternative virtue are socio-political. —Longino (1995b: 383)

...if one takes the primary objective of the feminist critique of science to be a restoration of what has been denied by modern science, then it necessarily has to explore the relationship between the world of science and the world it marginalises...It is within this framework that the post-colonial feminist begins to address issues both of modernity and of tradition... –Chadha (1996: 792)

This chapter, located within the philosophy of science and feminist engagements therein, draws upon Longino's reformulation of empiricism as contextual empiricism. It explores the implications of such a community-oriented epistemology from a location such as India, where tradition and modernity merge. Its focus is on developing the notion of ontological heterogeneity and the possibility of inserting local values in global interactions. Accordingly, the first part of the chapter presents the feminist

---

A draft of this paper was presented at the 'Gender and Science' Workshop collaboratively organized by the Indian Association for Women's Studies and the Research Centre for Women's Studies, S.N.D.T. University, Mumbai, 16–17 February 2010. Many thanks to Gita Chadha for her encouragement, discussion and insights without which this paper would not have been possible. All its limitations and flaws, however, are exclusively mine.

critique of empiricism. The second proceeds to explore prospects opened up by Longino's contextual empiricism. The final part examines the efficacy of contextual empiricism from a postcolonial feminist perspective.

Feminist Critique of Empiricism

Longino (1987: 52) observes, feminist critiques of science raise two types of questions: social and conceptual. The former concerns the social conditions under which misogynist or feminist science becomes possible. The conceptual question addresses the link between social standpoints or contexts and impersonal, objective, value-free inquiry. Over the past two decades, the methodological and epistemological aspects of science have witnessed a debate concerning the value-ladenness of scientific theory and whether this leads to epistemological relativism. As Jane Braaten (1992: 243) observes, numerous publications between 1984 and 1986 initiated this debate. She rightly cites Anne Fausto-Sterling's *Myths of Gender*, Ruth Bleier's *Science and Gender*, Evelyn Fox Keller's *Reflections on Gender and Science*, Sandra Harding's *The Science Question in Feminism* (1986), and articles by Ruth Hubbard, Helen Longino, Ruth Doell, Hilary Rose, and Susan Bordo as significant contributors. Philosophers adopting a feminist perspective have exposed the androcentric premises that underlie traditional approaches to philosophical knowledge. Feminist philosophers of science, such as Harding, Longino and Haraway have argued that mainstream philosophy's attitude to knowledge is characterized by dualisms of mind/body, reason/passion, nature/culture. To this Haraway adds human/animal, organism/machine, physical/nonphysical (1990).

Though at first sight this coalescing of diverse philosophical traditions as dualistic appears problematic, a closer look helps to understand the needs and claims of feminist philosophies of science. Harding traces the equation of masculinity with disembodied cognition to the specialist vocation of modern science. Arguably, modern science differs from its Greek counterpart being devoted to the experimental method that requires labour, rather than intellectual intuition.[1] But as Harding notes, modern science requires both the intellect and the labour of experimentation. She traces the lineage of modern science to the pioneers of modern technology– mariners, miners, foundrymen, carpenters, shipbuilders and the like who cut across the division of labour between intellect and physical work to make room for Galilean science. Harding points out that the questioning of authority of the ancients such as Aristotle and Ptolemy made personal experience available as the basis of knowledge. Besides critical scrutiny of old world views, openness to the future, non-prejudiced learning and a commitment to public service[2] all constituted the outlook of what Harding calls the new science movement. This outlook entailed a democratic and participatory ethos. Yet with the restoration in 1660 of Charles II's absolute rule in England, Harding observes that the demise of the Puritan Revolution of Oliver Cromwell was followed by the separation between the intellectual and social goals of science. Instead of being a part of society, science went on to occupy a niche position as a specialist discipline through the Royal Society in London, founded in 1662. It then became a purely cognitive endeavour that established itself as 'value-neutral', atomistic and experimental.

Further, in the late nineteenth and twentieth centuries, positivism in science revived Plato's theoretical enterprise (Habermas 1971: 302–14). Linking positivism to Plato elicits surprise because Plato was a critic of empiricism and gave no place to experimentation in his scientific methodology.[3] Habermas draws Platonism and positivism together as positions that are committed to a structure of reality independent of the knower. This parallel can be extended further to their asocial model of knowledge grounded in the idea that observations[4] are made by an atomic individual. Moreover, both approaches drive a wedge between analytic and empirical truths to privilege mathematics in a hierarchy of knowledge. Plato's 'Divided Line' (1987a), underscoring mathematics as a superior form of knowledge and the Carnap's thesis (1934) regarding 'unity of science' with physics occupying a position of supremacy, are two sides of the coin of an asocial approach to truth. Hilary Rose (1987: 153) aptly terms this approach as internalism, where science is seen as '...a story of cognitive transformation in the independent development of intellectual structures'. As far back as 1983,

Harding and Hintikka (1983: ix) declared that feminist intervention in science 'must root out sexist distortions and perversions in epistemology, metaphysics, methodology and the philosophy of science-in the "hard core" of abstract reasoning thought most immune to infiltration by social values'.

Yet critics such as Cassandra Pinnick (1994; 2005) have debated connecting the 'hard core of science' with society, although Pinnick (2005) has no problem with Longino's first feminist question regarding critiquing the social context of asymmetrical gender relations so that science practice becomes egalitarian. She notes that it is not peculiarly feminist to critique prejudice. The history of philosophy of science according to Pinnick (1994: 653) does not support the thesis that women have a unique control over removing sexism in science. Yet on the ground of social context, it is problematic to grant epistemic privilege to women in the name of their marginalization. Pinnick (2005) observes that along with other philosophers of science Longino has to demonstrate that the inclusion of those who are marginalized improves the very notion of scientific rationality and understanding. Hence, Pinnick (1994: 646) believes that Longino's second question that the 'hard core' of scientific knowledge and methods is androcentric cannot be taken 'seriously'. She argues that making scientific methodology social will generate a relativistic variety of positions so that the validity of feminist methodology cannot be maintained.[5] To resolve this conflict, Pinnick (1994: 650–51) suggests the turn to empiricism. Pinnick accuses feminist philosophers of science of appealing to analytical philosopher Willard Quine (1953) and Thomas Kuhn (1970) to endorse their alleged non-cognitive agenda in science. (Quine argued that empirical knowledge cannot be divided from its logical or non-empirical counterpart, and that experience is never immediate but always mediated through language, society, theory and the like; Kuhn turned to history to understand science and held that scientific discoveries are made within frameworks or paradigms.) However, rather than advocate the irrational, Harding and Longino amongst others turn to Quine to comprehend experience as theory-laden, rather than immediate and offer an alternate way to conceptualize the concept of experience.[6] Their internal critique of empiricism[7] and Longino's reconstruction of empiricism have offered more inclusive methodologies which have expanded the notion of scientific reasoning in conceding that the gender of the researcher affects the object of knowledge.

In their internal critique of science, Harding and Longino argue against immediate experience as the bedrock of science because experience cannot be sustained without recourse to language, culture and society. Harding (1986: 36–37) observes that Quine's attack on 'the two dogmas of empiricism' challenges the view that science and society are not interrelated.[8] Since language plays a role in the articulation of scientific facts and experience, the latter are inextricably bound to society. It is when the link between science and society becomes clear that gender can enter into the discourse of science. The feminist critique of the assumptions of value-neutrality, asociality, individualism and atomism in modern science is based on the relation between science and society. This critique focuses on how the very notion of asocial truth reflects the assumptions of a patriarchal society.

Thus, the feminist criticism against empiricism reveals that the so-called value neutrality underlying conventional science 'defends and legitimates the institutions and practices through which powerful groups can gain the information and explanations that they need to advance their priorities' (Harding 1992a: 568). But this is not what Pinnick (1994: 648) calls a license for undemocratic and even genocidal practices in science. Her thesis is that such a call for feminist intervention in science would take away the latter's objectivity in the course of politicizing it. Yet the point made by Harding is the reverse that if one were not aware of the biases that underlie science in the guise of value neutrality one might endorse precisely the kind of social engineering that Pinnick abhors.

Pinnick (1994) maintains that Harding rests her case against empiricism on the basis of her own view that women have a standpoint that needs to be introduced in science. However, this argument ignores Harding's internal critique of empiricism which has affinities with Longino who is a critic of standpoint theory. As Harding (1992b: 444) claims, feminist epistemology 'sets the relationship between knowledge and politics at the centre of its account'. Further, contrary to Pinnick's claims (1994: 647–49, 655), she does not renounce the need for objectivity, impartiality or critical thinking in science. Indeed, her point is that traditional empiricism that is founded on the observations of the knowing subject fails to acknowledge that observations are mediated by social relations. Harding has, thus, 'socialized' knowledge and shown that the stress on an individual knower outside society cannot attain the ideal of objective knowledge. Pinnick does not pay heed to this internal critique of empiricism. She also does not examine Harding's challenge to the notion of science as 'prediction'. According to Harding, instead of predicting and controlling the universe, science aims at opening the universe to a more interactive relationship with human beings.

Her critique of conventional empiricism notwithstanding (which she shares with Longino), Harding has not delved into its reconstructions that have been offered by Longino. In this sense, she assumes that her own standpoint theory provides the best alternative to the conundrum of androcentric science. According to Harding, feminist empiricism merely adds women to the traditional procedure of verification of data by an observer.[9] She argues that the feminist empiricists believed the possibility of eliminating androcentrism in science by following existing methods and norms of research; a view that is advocated by critics of feminist science such as Pinnick. As Harding observes, Marcia Millman and Rosabeth Moss Kanter maintained that the women's movement offers an enlarged perspective and provides for women scientists so that the androcentric bias in science can be corrected (1986). Yet as this does not question the methodology of science, empiricism takes the observer's experience as the data of science. However, Harding argues that considering that the observer is a gendered subject who occupies a social location, prejudice cannot be addressed by merely urging the observer to adopt an enlarged perspective. Harding's point is that androcentrism cannot be resolved by turning to standard empirical methods because it assumes the point of view of a disembodied thinker. Thus, she proposes her own theory of a women's standpoint as a solution to the problems plaguing feminist empiricism (which in Harding's view also applies to Longino).

As Longino (1987: 51) points out, an appeal to a 'women's standpoint' or a 'feminist standpoint' has a false universality, since women are too diversely situated to be coalesced into one group.[10] Thus, as Logino argues, the commonalities between a white working class peasant woman in Scotland and a Myanmarese peasant woman need to be spelt out by Harding (1993: 211). The basic problem with standpoint theory is that despite conceding to value-embeddedness of science and the problems with empiricism, as well as differences amongst women, it does not interrogate the specific ways in which all of these can enter into a feminist standpoint. Innumerable examples can be given to show that this affects the content and methodology of scientific research. This is precisely why there is a need to reinvent empiricism in the feminist context. That science is not a value-free enterprise is evident in its practice.[11] Moreover, Longino maintains that this is because it is a communitarian activity. The theoretical assumptions that surround scientific inquiry are chosen by inquirers who belong to a community of scientists and of society.

Longino (1987: 53) suggests that science be viewed as practice instead of content, as a process and not an end commodity. Thus, instead of feminist science, there should be a turn to 'doing science as a feminist ... The doing of science involves many practices: how one structures a laboratory (hierarchically or collectively), how one relates to other scientists (competitively or cooperatively), how and whether one engages in political struggles over affirmative action. It extends also to intellectual practices, to the activities of scientific inquiry, such as observation and reasoning.' The following section turns to Longino's specific reconstruction of empiricism as contextual empiricism.

Contextual Empiricism

Harding's basic discomfort with feminist empiricism is that it merely adds the word 'woman' onto existing patriarchal methods, without calling for an investigation into the historical character of the method itself. However, Longino's contextual empiricism offers a remedy to this problem. She maintains that philosophers tend to underplay the social, while sociologists emphasize it.[12] In order to bridge the divide between the two she corrects their common assumption that the social dimension is a source of prejudice. For Longino, the social dimension can provide a frame of rationality to arrive at knowledge that is not biased against women. She attempts to integrate the social with the logical to work out a contextual perspective of scientific knowledge with an inter-subjective basis for justifying scientific belief.

Longino (ibid.: 54) distinguishes between constitutive and contextual values in scientific inquiry. The former are shared assumptions through which scientific researchers become members of the scientific community. In contrast, what she terms as contextual values are social, political and cultural assumptions or the context of discovery that are normally not taken into account by scientists. She argues that the mainstream philosophical account of modern science as having value freedom assumes that these two sets of values are unrelated to each other. Longino, thus, provides a reconstruction of empiricism by taking contextual values into consideration. She argues that experience cannot justify knowledge claims directly. Citing logical positivism as an instance of the inability to negotiate between theoretical and observational language, Longino (ibid.: 54–55) offers an example: in particle physics terms like 'electron', 'pion', 'muon',' 'electron spin', and the like are used but it does not formulate a hypothesis such as: 'a pion decays sequentially into a muon, then a positron' by directly observing pions, muons and positrons. Rather than direct experience, observation is made up of photographs comprised of lines. In formulating data on their basis, the particle physicist assimilates 'theoretical and observational moments'.[13] Longino's interface between contextual and constitutive values can be extended to the biological sciences. A critical look at reproductive technology reveals the connection between, for example, the ideology of motherhood and the 'facts' that are presented as evidence.[14] The paradigm of modern science reproduces the patriarchal vision of a woman as a 'machine' who copies and nurtures the male figure/the father. Evelyn Fox Keller (1992: 42) aptly remarks that biological sciences that pursue 'the secret of life' (of course, here women's secrets are extracted) are suffused with metaphors of death so that these sciences are 'not life-threatening but life-less'.

Thus, according to Longino, experience can count as evidence for hypotheses when it is mediated by theoretical and social assumptions. She maintains that theory and data are not stable entities but that they change with time. Logical positivists viewed scientific inference as the outcome of hypothetical evidence and assumed that there is a formal connection between data and hypotheses. However, this formal structure cannot be transposed to the sciences because hypotheses are not just generalizations of data statements. Rather than a formal relation, data and hypotheses are connected through substantive assumptions regarding the field in which they operate and which the scientist theorizes. Theories are confirmed through the hypotheses that constitute them. These assumptions are not spelt out very clearly for the most part—yet they become 'the vehicle for the involvement of considerations motivated primarily by contextual values' (Longino 1987: 55). Background assumptions operate at many levels and their role is discerned and controlled through interaction among scientists. As a result, criticisms must proceed from a plurality of perspectives, which Longino upholds, and can be plenty. Longino (ibid.: 54) also evokes Quine's thesis of under-determination, which claims that any theory is always 'under-determined' by evidence that is offered in its support, so that diverse and even incompatible theories can have the same body of evidence.

Longino provides a list of six tentative features that emerge in the practice of science that take both context and justification into account and will also help in the development of a robust feminist science critique (1993: 336–38).[15] These include:

1. *Empirical Adequacy*: Citing feminist scientists who have exposed the limitations underlying so-called value-neutral programmes and androcentric research,[16] Longino mentions inclusiveness of race, class and gender as benchmarks of empirical adequacy. Such inclusiveness is necessary, but not sufficient for doing science as a feminist.
2. *Novelty*: Longino points out that the science that emerges as an alternative to its androcentric counterpart should have

a newness to it. She sees Harding's 'successor science' and Donna Haraway (1992: 336) appeal to science fiction writers as instances of such novelty.
3. *Ontological heterogeneity*: For Longino, doing science as a feminist will have to be sensitive to differences in the data that are studied. She cites Barbara McClintock's work on individual kernels of corn to discern mutability. Rather than homogenize differences from the point of view of white male superiority, Longino upholds the view that 'difference is resource, not failure'.
4. *Complexity of relationship*: Interaction between the diverse components within a framework that embraces both contextual and constitutive values is complex. Hence, according to Longino complexity has to be taken as the basis for explanation. She refers to Keller, *The Life and Work of Barbara McClintock* and her own interactionism as examples in this context.
5. *Applicability to current human needs*: Longino emphasizes that the purpose of scientific understanding is to improve the human condition or remove some of its suffering. Rather than militaristic control, 'scientific inquiry directed at reducing hunger (by improving techniques of sustainable agriculture, soil preservation, and so on), promoting health, assisting the infirm, protecting or reversing the destruction of the environment'.
6. *Diffusion of power*: For Longino, decentralization is more suited to feminist ways of knowing. She points out that 'one gives preference to research programs that do not require arcane expertise, expensive equipment, or that otherwise limit access to utilization and participation'.

Each of the theoretical virtues defined by Longino reveal that scientific discoveries take place within social practices and communities. Rather than represent an immutable reality, science endeavours to remedy misery. On the basis of an array of empirical evidence,[17] Longino distinguishes between reductionist linear methods and non-reductionist ones. She shows that what counts as evidence varies in important ways, depending upon the prior commitments of the inquirer to a background of values, which, in some cases are sexist and not in others. However, as Braaten (1992: 162) observes, Longino does not label such research as 'bad science' because no 'set of data is evidence for a hypothesis independently of some background assumption(s) in light of which the data acquire evidential relevance' (162). That evidence takes shape in the context of a set of background assumptions does not take away its worth as evidence. As Braaten (ibid.: 244–45) points out 'In the criteria of good and bad science, the constitutive values themselves involve intricate commitments to certain contextual values.'[18] According to Longino (1987: 61), since knowledge is moulded by values that are found in culture and, moreover, since one can choose one's culture, it is 'clear that as scientists/theorists we have choice' and can either do conventional science or contextual science with a commitment 'to traditional establishment or to our political comrades'.

Longino (ibid.) gives a new lease of life to empiricism by acknowledging observational evidence as theory-laden. Instead of privileging the standpoint of women with its homogenizing tendencies, she dwells on the subtle process of scientific inquiry that combines context with logic. By shifting knowledge to the structure of the community rather than the individual inquirer, Longino (1995a; b) allows for criticism, response, public evaluation and equal intellectual authority of researchers as a part of the scientific enterprise. Thus, in her attempt to integrate the contextual and constitutive domains of science, she makes space for both logic and holism.[19] She maintains that though positivists neglect history, their rooting confirmation of any generalization in stable evidence or constitutive values is 'logically satisfactory'. Alternatively, though holism takes history seriously and provides flexible contextual values, it is not satisfactory from a logical point of view, since it allows for variations in what counts as evidence. Longino tries to make room for both logic and holism by arguing that 'the culture of science is at once a part of the larger culture, participating in the formation of its contextual values, while at the same time a community committed to particular values constitutive of scientific research' (Braaten 1992).[20] Further, she suggests that an interaction between context and justification will lead to local knowledge (Longino 1987: 62).

All of this would help gesture, she argues, towards a science that is more inclusive and community-based, which is not geared towards commercialization and militarism.

Longino's fallible account of theoretical virtues that frame scientific inquiry allows for relating practices in the laboratory with the larger social world. It shows how, for instance, by introducing difference in one's research methodology, such a relation shapes the very notion of method. Yet Longino's project has the possibility of being open to Pinnick's criticism in replacing a consensus based on evidence with that of community, given her focus on decentralization, which concedes too much to the local. It conflicts with her theoretical virtue of empirical adequacy as it opens itself to the danger of doing science from a patriarchal perspective. The following section discusses this possible problem in Longino's turn to the local.

Choice, Community and Localism: Postcolonial Feminist Apprehensions

Longino (1987: 59) upholds the view that scientists adopt a specific paradigm of inquiry, such as reductionist or holistic, through choice influenced by contextual values. Thus, the scientific inquirer's relation to data emerges from a decision to adopt either the value-neutral linear or the value-laden interactive model prior to investigation. Longino cites her own research with Ruth Doell on neuro-science as a case in point (ibid.). Value-neutral neuroscience bases itself on sex-gender dualism to demonstrate how hormone exposure determines gendered roles. Against such determinism, Longino's and Doell's research assumes that human beings have the capacity for self-consciousness, self-reflection and self-determination. It focuses on the human brain's ability to reflect and express these self-determining capacities, by treating intentionality as central.[21] Thus, the 'political commitment' to feminism is expressed in part through the choice of the scheme of interpretation that is adopted by Longino and Doell. Longino (ibid.: 61) argues 'for the deliberate and active choice of an interpretive model and for the legitimacy of basing that choice on political considerations'. The virtues of ontological heterogeneity, complexity and diffusion of power are significant in such an inquiry which requires that the scientific community choose contextual values and where the choice is determined through prior commitment. Longino approvingly cites Harding for extending the feminist critique and reconstruction of science in a postcolonial context. Harding proposes 'postcolonial cross-cultural studies of science, where she makes a plea for the value of local knowledge systems that are themselves complicated by gender and class experiences.'[22]

Linking choice with the contextual values upheld by a community, however, does not necessarily entail doing science as a feminist. Arguably, the community is central while engaging with science in its several senses, such as understanding nature or as institutionalized practices.[23] Feminist science, located both in the developed and the developing world, has retained the relation between these various imports of science in their turn to local communities.[24] Moreover, feminists have repeatedly turned to the interaction with the community as the basis of scientific research. Several feminists have also applauded local knowledge as crucial to the enterprise of feminist science. Yet local communities are neither homogeneous nor feminist. There is a plurality of conceptual schemes operative at the local level. Leaving feminist research to choice would allow those who do not choose to be politically committed to feminism to persist in pursuing value-neutral science. Longino herself admits androcentric values are available in contexts of scientific research. Further, Longino advocates interspersing the local with global perspectives in her observation that 'ecofeminists and feminists in developing regions urge the development of technologies that are accessible and locally implementable. Some implementations of computer technology are valued for their ability to connect different but highly specific sites in widely spread, potentially global communication networks.'[25] Yet inserting local interests in global contexts does not guarantee the practice of science as a feminist. This problem is unique to feminist critiques of modern science that emerge from postcolonial locations.

These contradictions afflicting Longino can be illustrated by considering the problem of female foeticide in contemporary India, which is responsible for imbalanced sex ratios between males and females.[26] This problem brings together global technologies with the local androcentric value of preference for the male child. Hence, it uses the technology of prenatal diagnosis to determine the sex of the foetus. Prenatal diagnosis itself was introduced in India to facilitate population control and permit the birth of so-called healthy babies. Prenatal diagnosis is a modern scientific invention; it attempts to know the human body through privileging of the gaze of the technologist. Thus, the objectifying technology of prenatal diagnosis—a constitutive value, which is also rooted in the neocolonial interests of privileged western nations—thrives through local prejudice.

Prenatal diagnosis was first introduced in India in the mid-1970s for family planning and identification of so-called 'birth defects' (Menon 2004). Although initially, this technology was used against differently abled children, it subsequently lent itself to the identification of female foetuses.[27] But when women's groups protested against its role in sex-determination tests, this was declared illegal. Indeed as Menon (ibid.) observes, the Medical Termination of Pregnancy Act, 1971, that was introduced to legalize abortion in India was oriented towards family planning, health and eugenics Unlike the West, India does not have a strident debate on abortion because it was justified as a way of controlling women's bodies through family planning (ibid.).[28] Thus, rather than value-neutrality, the technology of prenatal diagnosis embodies the controlling dimension of gaze; moreover, it also has the aspect of discrimination built into its operation. Both instances exemplify an easy integration of local social intolerance towards the differently abled and women.

As Stanworth (1987) observes, one of the problems with technologies such as prenatal diagnoses is that it is expensive. Further, as she suggests, the expenses incurred for developing screening techniques to identify diseases can be channelized towards cure and care for those very diseases. Moreover, the social side of health and medicine is completely ignored with an emphasis on genetics. The local value itself treats biology as the basis of human identity. Inherited traits and genes are considered primary so that adults and foetuses are screened for 'genetic disorders' (ibid.: 22). Genetic parenthood stresses the origins of the child in blood ties, rather than the day-to-day child raising. Such an assumption of the primacy of blood ties and the inheritance of genes is prevalent at the local level both among men and women in India.[29] Many policy-makers openly state that they are ridding the world of having to provide for 'disabled' persons and sparing the public tax burdens.[30] In the western world itself, screening of the foetus came into existence under highly suspect local values of eugenics directed against the differently abled and non-white races (ibid.: 28–29).

Given the complexities of local knowledge, as well as the merging of global technology with local communities, one cannot assume as Longino does that the very notion of local is feminist and emancipatory.[31] The latter is assumed by her thesis of replacing the individual inquirer by the community (1993: 334–35). Janet Kourany suggests that rather local values and practices have to be filtered through the feminist critique, so that there are ' many voices' (1998) while doing science from a feminist perspective. Moreover, feminist science cannot be left to individual choice, but would have to be taken as the necessary condition of the laboratory itself. In this respect, Lisa Weasel (2001) suggests a more effective model for incorporating the local from a feminist point of view on the basis of the model of 'science shops' that emerged in Netherlands in the 1970s. Science shops were set up by university research centres as ways of connecting communities with laboratories. Students supervised by professors managed these shops which provided support and took input from local communities. Although the research in such a model is interactive, the feminist activist perspective can be brought in to examine possible androcentrism, inequalities of power and exclusions. Rather than simply take modern science and insert it in a local context, such a triadic relationship will enable rethinking the very assumptions of such technology to explore alternative prospects. Thus, rather than equate feminist consciousness with the practices of the community, it is understood as a distinct partner with the community and laboratory in producing research that is gender sensitive. The activities of Women's Studies Departments and the feminist critique of prenatal diagnosis in India are steps towards critically scrutinizing the local, rather than valorizing it.

Notes

1. This point has been made by both non-feminists such as Michael Foster (1934; 1936) and feminists such as Sandra Harding (1986). Further, modern science's disembodied masculinity dovetails with its Greek predecessor. The discussion of the origin of modern science is derived from Harding (ibid.: 179–215).
2. Galileo believed in science for the people as Harding observes. Is there a source from Galileo too? Also see Brecht's play *Galileo* (1960).
3. Plato's dialogue '*Theatetus*' is a case in point (1987b), see Habermas (1971). There are other differences between the ancient Greek approach to science and the universe, in relation to the modern. The latter emerged in sixteenth-century Europe and the findings of its key scientists such as Copernicus, Bruno, Kepler, Galileo and later Newton were canonized into a scientific method by modern philosophers such as Hobbes, Descartes, Locke and others.
4. Observation varies from the non-sensuous to the sensuous from Plato to positivism.
5. Pinnick cites Sandra Harding whose work she critiques in order to prove her thesis that this contradiction implies that feminist epistemology does not work. Harding has extended feminist theory to epistemology and natural sciences.
6. Pinnick herself notes that the project of feminist epistemology has been inspired by the critique of science through works such as Kuhn's *Structure of Scientific Revolutions* (1970) and Quine's under-determination thesis, the Strong Programme in the sociology of scientific knowledge, and general themes within the feminist critique of modern society (Pinnick 1994). But Harding, Haraway and Longino, have all suggested ways of developing an epistemology with greater links with society.
7. Sandra Harding has developed this critique in detail (1986).
8. According to Harding, '[T]he problem with the conventional conception of objectivity is not that it is too rigorous or too "objectifying," … but that it is not rigorous or objectifying enough….' (1992b: 438).
9. Harding's views on empiricism are derived from her account of the same and its contrast with feminist standpoint theory (see Harding 1986: 136–62).

10 Quoting Maria Lugones and Elizabeth Spelman, Longino (1987: 53) argues that the diversity of women's voices cannot be reduced to a single cognitive framework that homogenizes them.
11 See Mahadevan, 2008 for a critique of the alternative between value-neutral knowledge and standpoint theory.
12 Pinnick herself questions this distinction as 'sheer caricature', as philosophers have been debating the arational dimension in science (2005). Yet philosophers such as Husserl or Heidegger have not turned to the social dimension in their respective criticisms of science. Indeed, even metaphysically oriented philosophers have indicted science for not adopting a larger world view and the like. With the exception of the Marxist tradition, philosophical critiques of science have not tried to recover the social dimension.
13 'A skeptic would have to be supplied a complicated argument linking the elements of the photograph to traces left by particles and these to particles themselves.' (Longino 1987: 55).
14 This argument is largely derived from Michelle Stanworth (1987). Also see Maria Mies (1993) and Luce Irigaray (1985) for similar critiques.
15 Longino qualifies these as follows: 'The list of six I propose to discuss should be understood as a sample, rather than as a definitive set. Nevertheless, I believe it has features that would characterize any such list. In particular, I think any list will contain, as this one does, formal, substantive, and social or practical elements' (1993: 336).
16 Her examples include Ruth Bleier, Anne Fausto-Sterling, Richard Lewontin, Ruth Doell.
17 Even critics such as Pinnick acknowledge that Longino has contributed both to philosophy, history of science and science practice.
18 According to Pinnick, Longino avoids relativism by taking recourse to normative sociology rather than epistemology.
19 This argument is conducted in considerable detail, beginning with a general discussion of the debate in the philosophy of science in the last half century, between (among others) the neo-positivists Carl Hempel and Karl Popper, and their opponents, the 'holists' Thomas Kuhn and Paul Feyerabend.
20 To develop her case, Longino provides instances from biology, physical anthropology, physiological and psychological research on sex

differences in brain organization, and neuro-endocrinological studies of the effects of sex hormones on behaviour.
21 Pinnick discusses an instance that contributes in a positive way to linking social interaction with science, namely, Wray (2002). She points out that Wray attempts to see the way collaboration has a causal epistemic effect on scientific data. It allows for pooling together limited resources of research in ways that improve upon non-collaborating individual scientists. Scientific research teams successfully reach their epistemic goals after carrying out research through collaboration, unlike those who do not collaborate. Pinnick quotes Wray as saying that 'Ideally, it would be useful to have information on specific research groups, showing increased productivity after collaboration, followed by greater funding, which in turn would be followed by continued collaboration. Unfortunately, at present, such data are not available' (158–59). Thus, Pinnick points out that though Wray is able to trace the social trend of collaboration in the practice of science, he is not able to show productivity as its epistemological consequence. However, in citing this example of inconclusiveness, Pinnick ignores the various conclusive instances in neurobiology, that have been researched, for example, by Longino.
22 Harding as cited in Kohlstedt and Longino (1997).
23 Following Kohlstedt and Longino, sciences are varied and constituted by a variety of linguistics uses and practices (1997).
24 See, for instance, Harding 2001; Manorama and Walters 2001; Mies 1993; Weasel 2001.
25 However, studies such as that of Kelkar (1999) and Schiebinger (2004) are needed to assess the patriarchal connotations of local knowledge both within the culture and across the globe.
26 The term 'female foeticide' has been retained to convey the injurious and discriminatory character of such abortions. Feminists have suggested that this expression be replaced by 'sex selective abortion of female foetuses' (Menon 2012). For the term 'foeticide' with its connotation of murder has been used by the American right in their discourse against abortion. However, words do not have an inextricable link with the contexts in which they emerge. Moreover, those who engage in this practice do so because they discriminate against women. Their intent is eugenics and not abortion. Thus, one can use the term 'female foeticide' and yet advocate abortion from a feminist point of view.

27 See Ghai and Johri (2015) for a discussion of these themes.
28 Although many local knowledge systems are against abortion, not all of them do. Londa Schiebinger has done a study of Amer-indian medicines to show how local medicine did have techniques for abortion which were used amongst them (2004). Moreover, she also points out that even though the colonizing inquirers–some of who were women–knew about it, they did not circulate this knowledge in the West because they were committed to women's roles as reproducers.
29 Writing in the western context, Stanworth reveals that surveys indicate that such technologies are choices that men would make rather than women, who would prefer adoption (Stanworth 1987: 22); also see Rose 1987 for a critique of reproductive technologies from a feminist perspective.
30 See Kelkar (1999) for a critique of Ayurveda from a feminist point of view (1999).
31 See also Mies (1993) for this assumption.

References

Braaten, Jane. 1992. 'Reviews of *The Politics of Women's Biology* by Ruth Hubbard, *Science as Social Knowledge* by Helen Longino; *Body/Politics. Women and the Discourses of Science* by Mary Jacobus, Evelyn Fox Keller, and Sally Shuttleworth'. *NWSA Journal* 4, 2: 240–47.

Brecht, Bertolt. 1960 [1955]. *The Life of Galileo*. Madras: Oxford University Press.

Carnap, Rudolf. 1934. *The Unity of Science*. London: Kegan Paul.

Chadha, Gita. 1996. 'The Science Question in Post-Colonial Feminism', *Economic and Political Weekly* (henceforth *EPW*) 32, 15: 791–92.

Code, Lorraine. 1998. 'Voice and Voicelessness: A Modest Proposal?' In *Philosophy in a Feminist Voice: Critiques and Reconstructions*, edited by Janet A Kourany. Princeton, NJ: Princeton University Press: 204–30.

Ghai, Anita and Rachana Johri, 2015. 'Science, Gender and Reproductive Technologies: A Case of Disability', *Feminists and Science*, vol. 1. Kolkata: Stree: 96–121.

Foster, Michael. 1936. 'Christian Theology and Modern Science of Nature', *Mind* 45, 177: 1–27.

———. 1934. 'The Christian Doctrine of Creation and the Rise of Modern Science', *Mind* 43, 172: 446–68.

Habermas, Jürgen. 1971. (1968). *Knowledge and Human Interests.* Boston: Beacon Press.

Haraway Donna.1990. 'A Manifesto for Cyborgs: Science, Technology, a Socialist Feminism in the 1980s'. In *Feminism/Postmodernism*, edited by Linda Nicholson. New York: Routledge: 190–233.

Harding, Sandra, and Merrill B. Hintikka.1983. *Discovering Reality: Feminist Perspectives on Epistemology, Metaphysics, Methodology and Philosophy of Science.* Dordrecht: D. Reidel.

Harding, Sandra. 2001. 'After Absolute Neutrality', *Feminist Science Studies*, edited by Maralee Mayberry, Banu Subramaniam, Lisa H. Weasel, 291–304. New York: Routledge.

———. 1992a. 'After the Neutrality Ideal: Science, Politics, and "Strong Objectivity",' *Social Research* 59: 567–82.

———. 1992b. 'Rethinking Standpoint Epistemology: What Is "Strong 'Objectivity?"' *The Centennial Review* 36: 437–70. 656–57.

———. 1986. *The Science Question in Feminism.* Ithaca, N.Y.: Cornell University Press.

Irigaray, Luce. 1985. *Speculum of the Other Woman.* Ithaca. N.Y.: Cornell University Press.

Keller, Evelyn Fox.1992. *Secrets of Life/Secrets of Death: Essays on Language, Gender and Science.* New York: Routledge.

Kelkar, Meena.1999. 'Biomedicine, Ayurveda and Women', *Issues in Medical Ethics* 7.

Kohlstedt, Sally Gregory, and Helen Longino.1997. 'The Women, Gender, and Science Question: What Do Research on Women in Science and Research on Gender and Science Have to Do with Each Other?' *Osiris* 2nd Series, 12, Women, Gender, and Science: New Directions, Chicago: Chicago University Press: 3–15.

Kourany, Janet A. 1998. 'A New Program for Philosophy of Science'. In Many Voices'. In *Philosophy in a Feminist Voice: Critiques and Reconstructions*, edited by Janet A. Kourany. Princeton, NJ: Princeton University Press: 231–62.

Kuhn, Thomas. 1970 (1962). *The Structure of Scientific Revolutions*. Chicago: Chicago University Press.

Longino, Helen. 1995a. 'To See Feelingly: Reason, Passion, and Dialogue in Feminist Philosophy'. In *Feminism in the Academy*, edited by Donna Stanton and Abigail Stewart. Ann Arbor, MI: University of Michigan Press: 19–45.

———. 1995b. 'Gender, Politics, and the Theoretical Virtues', *Synthese* 104, No.3: 383–97.

———. 1993. 'Feminist Standpoint Theory and the Problems of Knowledge', *Signs* 19, 1: 201–12.

———. 1992. 'Taking Gender Seriously in Philosophy of Science'. In *PSA: Proceedings of the Biennial Meeting of the Philosophy of Science Association*, vol. 2: 333–40.

———. 1990. *Science as Social Knowledge: Values and Objectivity in Scientific Inquiry*. Princeton, N.J.: Princeton University Press.

———. 1987. 'Can There Be a Feminist Science?', *Hypatia* 2, 3: 51–64.

Mahadevan, Kanchana. 2008. 'Rorty, Haack and Feminist Epistemology', *Journal of Indian Council of Philosophical Research* 35, 2: 27–49.

Manorama, Swatija, and J.Elaine Walters. 2001. 'Fertile Futures: Grounding Feminist Science Studies Across Communities'. In *Feminist Science Studies*, edited by Maralee Mayberry, Banu Subramaniam, Lisa H.Weasel. New York: Routledge: 248–57.

Menon, Nivedita. 2004. *Recovering Subversion*. New Delhi: Permanent Black.

———. 2012. 'Abortion as a Feminist Issue: Who Decides and What?' Kafila http://kafila.org/2012/05/11/abortion-as-a-feminist-issue-who-decides-what. accessed on 14 February 2013.

Mies, Maria. 2010. [1993] 'New Reproductive Technologies: Sexist and Racist Implications'. In *Ecofeminism*, edited by Maria Mies and Vandana Shiva. Jaipur: Rawat: 174–97.

Pinnick, Cassandra L. 2005. 'The Failed Feminist Challenge to Fundamental Epistemology', *Science and Education* 14, 2: 103–16.

———. 1994. 'Feminist Epistemology: Implications for Philosophy of Science', *Philosophy of Science* 61, 4: 646–57.

Plato. 1987a. [1961]. *Republic*. In *The Collected Dialogues of Plato*, edited by Edith Hamilton and Huntingdon Cairns, 575–844. Princeton, NJ: Princeton University Press.

Plato. 1987b. [1961] *Theaetetus*. In *The Collected Dialogues of Plato*, edited by Edith Hamilton and Huntington Cairns, 845–919. Princeton, NJ: Princeton University Press.

Quine, W.V.O. 1953. 'Two Dogmas of Empiricism'. In *From a Logical Point of View*. Cambridge, MA: Harvard University Press: 20–46.

Rose, Hilary. 1987. 'Victorian Values in the Test-Tube: The Politics of Reproductive Science and Technology'. In *Reproductive Technologies: Gender, Motherhood and Medicine*, edited by Michelle Stanworth. Cambridge: Polity Press: 151–73.

———. 1983. 'Hand, Brain, and Heart: A Feminist Epistemology for the Natural Sciences,' *Signs* 9: 73–90.

Schiebinger, Londa. 2004. 'Feminist History of Colonial Science', *Hypatia* 19, 1: 233–54.

Stanworth, Michelle. 1987. 'Reproductive Technologies and the Deconstruction of Motherhood'. In *Reproductive Technologies: Gender, Motherhood and Medicine*, edited by Michelle Stanworth. Cambridge: Polity Press: 10–35.

Weasel, Lisa H. 2001. 'Laboratories Without Walls'. In *Feminist Science Studies*, edited by Maralee Mayberry, Banu Subramaniam and Lisha H. Weasel. New York: Routledge: 305–20.

Wray, K. Brad. 2002. 'The Epistemic Significance of Collaborative Research', *Philosophy of Science* 69, 1: 150–68.

15

Traditional Knowledge and Feminist Dilemmas: Experience of the Midwives of the Barber Caste in South Tamil Nadu

Meena Gopal

If there is an area that engages social science research and beckons state policy but has faced consistent neglect, it is that of caste-based occupations and their status as knowledge-making systems. A historically persistent feature of our social universe, caste-based occupations have relegated substantial sections of our society into socially necessary occupations that are consistently devalued and considered too degrading for others' involvement. This chapter emerges out of a synergetic moment when feminists in Maharashtra, engaged in spirited debates over women's role in caste-based occupations.[1] It aims to reflect and extend the debate into the sphere of the politics of knowledge making in a postcolonial society like India, reflecting the recent concerns of women's studies with critiquing and reinventing knowledge-making systems, including modern western science, from feminist perspectives.[2]

For decades, the caste-based occupations present in several rural and some urban areas had not sufficiently grabbed the attention of the women's movement in India. The ban imposed in 2005 by the government of Maharashtra on women dancing in dance bars drew diverse feminist responses, with some seeking revocation of the ban whereas others supported the ban, voicing opposition to the continuation of women's entry into caste-based occupations symbolized by the sexual labour of women dancing in dance bars.[3]

Following the impasse between the two camps, a dialogue was initiated with the intention of wading through the contradictions and conflicts within and between feminist groups, addressing the role caste plays in all labour and in the work done by women who come from communities that have traditionally performed certain tasks, and to seek ways of understanding feminist locations and politics. Several questions arose in the discussions (Gopal 2009), but the one that engaged a lot of attention demanding further exploration was the role of caste-based occupations and women's participation and in particular women's relationship to their work. The role of the female body, the attribution of stigma enveloping both caste and gendered dimensions, the histories that are enmeshed in the subjective engagements of caste feminists, dalit feminists and the women from communities who perform these labours, all of these were troubling issues that eluded a common resolution. While stigma, castigation, and deprivation of livelihoods were the experience of some of those engaged in caste-based labours, another set of concerns is the connection between stigma, denial, and appropriation of knowledge.

In the first section of the chapter, I discuss the relationship between caste hierarchies and knowledge generation, specifically tracing the social relations of groups that performed tasks related to the body. The gendered nature of these roles and the stigma attached to them also informs how exclusions operated within the social relations in a community. A brief historical review of one caste in a specific region demonstrates how feudal pre-colonial and colonial practices progressively sidelined the skills and social status of the barber caste who were traditional healers and *dais* (midwives). In the next section, I look at the experience of midwives from the barber caste in villages in south Tamil Nadu. Here, I trace the manner in which these midwives practiced their caste-based occupation earlier, the changes over time and their impact on the women as women and as practitioners of a caste-based occupation. The third section is an analysis of the manner in which modern institutions define the place of midwives within a healing system and as practitioners of a caste-based occupation. In conclusion, I reflect on how the Indian women's health movement engaged with women of the healing castes even as it grappled both with modernity and the state and societal control over women's health and their bodies. This also throws light on the dilemmas of feminists concerned with science and the production of knowledge, as also the experience of negotiating with bio-medicine and its interventionist potential that the women's health movement has striven to resist.

Caste and Knowledge Hierarchies

The relationship of knowledge systems and caste hierarchies has been a contentious matter in our social life ranging from historically exclusionary practices by upper castes denying access to formal learning to lower castes and communities, to colonial interventions that attempted to reverse or exacerbate these exclusions, and to present-day debates over state-initiated affirmative policies and action for educational reservations for historically disadvantaged groups listed in Article 15(4) of the Constitution. In India, knowledge generation and dissemination has traditionally been the preserve of the upper castes. This consolidation of the knowledge base through religious dictates which are made by the upper castes, and the power that ensues from this, created a social structure that excluded large sections of the lower castes and communities from access to knowledge (Chakravarti 2003). These exclusions and alienations did not go uncontested and spurred struggles in several parts of the country to secure access to education and the power and social capital that accrue from it (O'Hanlon 1985, Omvedt 1994, Geetha and Rajadurai 2008). In the colonial era, the British consistently encouraged the educated groups to be part of the colonial state; wherever there was elite opposition to the British, however, they encouraged education of the disadvantaged groups.

Thus far, the analysis of the relationship between caste and knowledge has acknowledged the exclusion of lower castes from traditional knowledge and the spread of modern knowledge through secular institutions during the colonial and postcolonial era. There is, however, another dimension of caste and knowledge that needs attention, and that is the denial and/or appropriation of knowledge generated by castes lower in the hierarchy, which emerges through a dialectical relationship with their work, by those who do not perform these labours. In India several caste groups practise traditional and artisanal occupations that are a source of knowledge and translate into action for service, upkeep and care of the community, such as sweepers, leather workers, potters, barbers, or peasants. But the knowledge of the lower castes comes to be denied a space, 'invisibilized' or wiped out in the formal power structures of knowledge generation, and thereby lost or appropriated without any due being given to them.

The logic of the caste hierarchy operates on the notion of purity and pollution of castes and their labours. Upper castes are deemed pure while lower down the hierarchy there is a lessening of purity and increase in pollution, with the labour itself acquiring purity and impurity and being 'written' on the external body of the person. Feminists have addressed the relationship between caste status and the gendered body and the notions of purity and pollution attached to it, while also bringing into the debates the embodied nature of work (Chakravarti 1996; Chowdhry 1996; Rege 1996). The dynamic between caste and the gendered body is characterized by the control of women's sexuality and fertility in terms of endogamous marriage and childbirth, the surveillance of widows, sexual violence directed at women to uphold or violate caste and community honour, and several other practices and actions that are non-uniformly directed at women across castes, marking their bodies. Any challenge to these gendered practices invites stigma and ostracism. Despite these constraints, the labour of women is invaluable as it keeps the social reproduction going without making any claims for the value generated; it is, therefore, harnessed at any cost.

Ilaiah (2007) mentions very graphically the hierarchies and exclusions of the contributions of castes in Indian society. In order to retain the social power that accrues from control over knowledge generation, the knowledge generated by upper castes is deemed to be of more value than the skill of the low caste worker. Inbuilt into this are hierarchies of the labour of the mind versus the labour of the body, thereby grading some knowledges as being valuable and desirable while others are despised and undignified. This pans out in a complex way with respect to science which, on the one hand is abstract, theoretical and esoteric, and on the other (as opposed to philosophy), dirties its hands in the laboratory but remains exclusionary. Despite the productive and sustaining labours and skills of the lower castes contributing to the upkeep of a community, this is despised and devalued in the hierarchies of ritual and labour leading to the obliteration and loss of valuable knowledge and skills of the lower castes. There is a caste-based and gendered relegation of such labours. All caste-based occupations, be it the manual scavenger, the leather worker, the agricultural labourer, the midwife (who is the focus of this chapter) and the entertainer-dancer, operate through the specificity of the sexual division of labour. Women in all these labours occupy a subordinate position within the sexual division of labour as it structures itself within family labour. All too often the labour of women within these caste-based occupations remains subsumed within family labour and is often invisible. Women's labour thus becomes more marginal and exploited compared to men's. Despite the practices of erasure, there is evidence that women who perform these occupations derive and attribute meanings to their work and that this may not be reflected in the community understanding of the work and knowledges. Within the caste locations, women also draw and construct meaning over their *changing* labour, with shifts in socio-economic life (Gopal 2009).

It is only when women begin to transgress the meanings generated by the community and try to actualize their own rationale and the meaning they derive from it that there are contradiction and conflict. This was evident in the case of the women bar dancers. What is distinct is the way women in these caste-based occupations negotiate their position by evolving subtle means of resistance and subversion. Examples of such negotiation are the dalit woman agricultural labourer who engages with life even as she mocks the upper caste norms for funeral rites. Or the 'de-notified tribal' bar dancer, who migrates to find a better life for herself and her family, thereby avoiding the female family labour of prostitution (Siddiqui 2003; Agarwal 2006)–by an Act of 1871, the British classified some communities as 'Criminal Tribes'; independent India 'de-notified' these tribes. Or the midwife, who is able to negotiate a space for herself, by applying her skills to treat minor ailments and provide preliminary aid in case of emergencies in pregnancies, while also being available to participate in the ritual ceremonies of the dominant groups in the village.

While at one level there is the desire to eschew the stigma on part of the social movements who see caste-based occupations like midwifery as the reinforcement of discriminatory social hierarchies and notions of pollution and purity, at another level there is also the need to look at these labours and occupations as constituting productive resources and skills, which have either been appropriated without recognition or denied their due in the system of knowledge production. As already mentioned, feminist scholarship has demonstrated how an upper-caste, male-centred and westernized knowledge production system has alienated knowledge produced or created by women, lower castes and other marginalized groups. The colonial state while neglecting the health needs of the people encouraged the castigation of local healers and midwives by the missionaries and white women professionals (Ram 1998). The colonial enterprise through its distortion, neglect and non-interference has played an active role in divesting or appropriating the knowledges and resources of healers and midwives. The castes that were serviced actively distanced themselves on the pretext of the dais' lack of hygiene and being superstitious.

One such caste is that of the barbers, known as *kudimagan* in Tamil Nadu, who perform not only the act of removing hair for people within a community as part of grooming and hygiene, but also participate in certain ritual acts during birth and death and other rites of passage within the community. The women from the community known as *maruthuvacci* (woman doctor/healer) also perform certain community activities; for instance in southern Tamil Nadu, they are the village midwives. A historical review of the caste in Tamil Nadu provides a picture of the transformation of a community of healers into a caste entrenched within the ritually polluting castes (Ragupathi 2006). While Thurston (1909) lists the barber caste as impure, Mencher (1978) refers to the barber and washer castes as semi-untouchable. The review traces a gradual shift in social position where the brahmins would accept the surgical and healing services of the barber caste but refuse to accept food from them due to the proscription of the legal scriptures. The social organization ensured that even among the healing castes, there were distinctions depending on which castes they served. Separate sub-groups within them performed the rites and services of upper, middle and lower castes; these distinctions were also separate and rigid. Historically in many parts of India, the dai was part of the *jajmani* system attached to a feudal household along with the other artisans in a community. The dai's work however was women's work, an extension of domestic labour and hence devalued (Chawla 2002). This undervaluation of the dais' traditional contribution was exacerbated during the colonial period with a continued devaluation of indigenous healers and practitioners generally.

Since the healing castes (*maruthuva jati*) depended on the daily household contribution of food and material sustenance from the village population, which was hardly enough for them to maintain themselves and their family, their livelihood bordered on the precarious (Raghupathi 2006.). The feudal system saw to it that the kudimagan stayed entrenched in his caste occupation and was not able to eschew it, almost bonding him to his caste occupation. However, the dominant castes were able to dislodge one poor kudimagan (village barber) and replace him with another, thus causing economic insecurity for the healers. In the late colonial and as well as the postcolonial phase, the dominant castes began to move out into the cities where they began to seek the services of 'trained and educated' healers, leaving the women and men of the healing castes without a livelihood. Thus, this historical review demonstrates that their skill and knowledge were circumscribed by the caste-based economic stranglehold of the feudal society which spilt over into the postcolonial era that gave the partakers of their services newer opportunities while leaving the servers high and dry. At present, men and women from the healing barber castes are only called upon by other castes in the village for certain rituals, or for bathing the dead body, shaving and cleaning the corpse so that the soul goes to the other world; such practices only reflect the motives of the upper castes who wish to keep the caste rituals alive without any material advantage to those performing it. In this manner the skills and knowledge of the barber community are sidelined, neglected, and thereby lost, while newer professionals are employed in their place.

Midwives of the Barber Caste

In Tirunelveli district of south Tamil Nadu, the social location of the barber caste is low but not as low as that of dalit castes. The dominant caste in the village, the nadar, still seek the ritual presence of both barber and washer castes for puberty rites, marriages, birth and death ceremonies. They are provided with some food and other token gifts in recognition of their ritual role. This does not mean that their caste status is elevated or their material situation is at par with that of the dominant caste. Some men in the barber community have set up hair-dressing shops, taking advantage of a growing monetized economy. Secular and modern institutions such as education, transport and communication facilities and the access to non-traditional jobs have also altered the socio-economic condition of many people lower in the caste hierarchy. This is also responsible for several youth of subsequent generations not pursuing the traditional caste-based occupation.

In many villages in Tirunelveli district, barbers continue to perform their caste-based occupations and run barber shops within the village or at junctions of villages. They do this while also seeking other work within the village. Sometimes the men are invited to cook at ritual events like marriages and other rites for the nadar community. The generation of women from the barber community who assisted as midwives at birth are old and no longer practice. There are no replacements for their skill in conducting deliveries, as younger women do not practice the caste-based occupation. During my research (see note 1) in Tirunelveli district, Chinnamalai and Badramati in their fifties from Kalloorani and Keelapavoor in Tenkasi taluk, told me that they learnt the skill of conducting deliveries mainly from their mothers-in-law. Indeed, they claim to have used this skill in supervising and orchestrating their own deliveries. Chinnamalai's seven children were delivered with instructions from her and the help of others; two of her children were born at home.

Today, although they do possess adequate skills, both the women refrain from utilizing it to conduct deliveries for fear that anything might go amiss and they might be blamed at a later juncture. Even when they or their mothers-in-law had performed these tasks over ten to fifteen years ago, they got meagre remuneration. The fees for each activity was specified and decided at the village level by the dominant castes. For instance, to cut the umbilical cord they would get 25 paise or six *pakka* (approximately 2 kilograms) paddy for the whole year from the village. On the occurrence of a death, both men and women from the barber community are asked to perform the removal of hair, and other cleaning rituals of the dead person.

While the women's primary skill was that of conducting deliveries, they also possessed several medicinal skills to cure minor ailments, and had knowledge of certain herbs and plants as remedies. As Chinnamalai says, even today she has the self-confidence of curing several ailments and the knowledge of several plants but she utilizes this only for those who seek her out. The constant downgrading by both upper castes who devalue their skills as well as the attitude of modern institutions that eschew traditional skills, comparing them to the superiority of technological interventions, have undermined the confidence of traditional healers.

Yet, people in the village are constantly streaming to Chinnamalai's house with little babies, or women come to her for immediate relief from minor ailments, and she attends to them all. Several of the herbs and plants that are used are grown in the backyard, access to which is easy and these serve as a panacea for several ailments. Additionally, she also has minor surgical skills. She exuded confidence as she described the process by which she extricated a piece of metal that had got embedded in a man's foot and had immobilized him for nearly a week. Chinnamalai had applied oil and deftly extracted the metal piece. She charted with pride the repertoire of the healing, surgical and delivery skills that she had. Pride and indebtedness crowned the skills that she learnt from her mother and mother-in-law who were well respected and sought after as traditional healers and midwives in the village. She continues to be humble despite possessing a wide range of knowledges and skills.

Anjanai-paati in her eighties from Pappakkulam in Ambasamudram taluk was more graphic in the elucidation of her skills: 'A ripe fruit will not stay long on the branch, it has to drop.' Recalling her younger days when she was always on call by households where women were in different stages of pregnancy, she said it was a skill to recognize when a woman was ready to deliver. The baby would be like a fruit ripened and ready to drop, while the woman's body was like a branch that could no longer hold the baby. But cynicism marked her observations of the pregnant women of today who would rush to the hospital for a surgical intervention or a Caesarean section to deliver the baby. Today's women had no patience, she noted, and modern hospitals induce this impatience. She observed that people's urgency to seek medical and surgical intervention was directly proportional to the rise of commercialized medicine reflected in the upsurge of hospitals and nursing homes. She herself relied more on spices and condiments to assist in the delivery process (rather than herbs) as spices were commonly available in the shops.

Another former midwife had attended a training programme organized by the government health services system, but her experience has not been subsequently of much use for the community as she did not practice in the community thereafter. The midwives' experience of the training was as seekers of new knowledge, where their own experience and skills were shelved and which were not called upon to be demonstrated during the training (see also Chapter 20).

Modern Institutions, the Community and Caste-Based Knowledge

In the modern economy, there have been attempts at integrating traditional occupations into cottage industries and industrial production, for instance, the efforts to accommodate leather artisans and workers, who pursued hereditary caste-based labour such as flaying animals, tanning hides and producing footwear (Prasad and Rajanikanth 1991; Nihila 1999). But when the markets and the export potential for the leather industry developed, the non-polluting processes such as trading and marketing, which were also higher in the value chain were taken over by the non-artisan upper castes. The low value and disagreeable tasks of flaying dead animals and tanning hides, which were also considered polluting and demeaning, continued to be done by the traditional leather workers, the chamars or chakkiliyans.

Thus on the one hand, there is the state and the industrial elite who appropriate the productive labour of artisans and lower-caste workers, but exclude them from the surplus generating enterprises, while on the other, there is the appropriation and displacement of knowledge and skills produced by these castes in the performance of their caste-based occupations.

Yet, by and large, the modern economy has not just failed to integrate caste-based occupations within it but instead has entrenched these—and this is illustrative of the upper castes and the state colluding to keep the caste system alive in the new productive economies. The Ministry of Railways has been the worst offender in perpetuating the system of manual scavenging with the open discharge of toilets requiring the excreta to be manually gathered and lifted off the tracks (Subramaniam 2010). Rehabilitation has to include de-stigmatization of the so called menial jobs via changes in recruitment patterns and policies. A complete technological overhaul of the sanitation system is required if members of the staff are not required to work in the degrading conditions, in addition to reservation in educational and employment opportunities, as well as a comprehensive rehabilitation package.

The interface of caste-based knowledge and skills with modern state institutions, specifically scientific conceptions of the skills of nurses and midwives, that also takes into account caste relations in the village and impinge on the status and opportunity provided by skills/knowledge in the possession of the women from the barber community, raises several questions. How did the modern state perceive the skills of the traditional midwives vis-à-vis the Auxiliary Nurse Midwives (ANMs)/nurses of the state medical system? What was the state's logic in training Traditional Birth Attendants (TBAs)? Did it take into account the caste locations of these women?

Badramati was one of the few from the barber caste who attended a training programme for traditional birth attendants organised by the Tenkasi taluk hospital in the mid-nineties. But she has hardly conducted a delivery since then. The interaction with the health system gave her lessons in what she could adopt from that system, of care and cleanliness, not that it added much to her own capacities. People from the village call her when they need advice but they do not expect her to actually conduct or supervise the delivery. She is still called upon for all the accompanying ceremonies and rituals, from the post-partum stage to the naming of the child, and given an appropriate ritual compensation. Women from the barber community have been confined to their role in the caste hierarchy, which is maintained in the village through the conduct of rituals, but their knowledge is put to use in a very peripheral sense. There is an emphatic pursuit of modern means to fulfil Nehru's vision and the dream of the framers of India's Constitution–of ridding the country of disease, poverty, ignorance and prejudice. One of the means of accomplishing this is through aiming for 100 per cent institutional deliveries to improve the Human Development Index through important social indicators, namely, the nature and conduct of births, and its impact on maternal and infant mortality (Dreze and Sen 1995). The doctor at the community health centre at Keelapavoor said that the Tamil Nadu government has not conducted training programmes for TBAs/midwives since 2000, ever since she had completed her training as a medical practitioner and been in service. What is encouraged now are institutional deliveries, she was emphatic.[4] One of the achievements lauded by the developmental state is the extraordinary manner in which Tamil Nadu attained its demographic transition and replacement level fertility rate (Swaminathan 1998). One of its correlates is attaining 90 per cent institutional deliveries which is a record of sorts (NFHS III, 2007). The role of traditional midwives in helping the state secure the targets of sterilization for this successful implementation of the family planning programme is not unknown. Thus, the traditional midwife although relegated by the state still exists on its fringes in a utilitarian capacity.

The modern state's attempts to train the dais, an exercise at implementing the World Health Organization's (Vederese and

Turnbull 1975) 'TBA retaining' strategy following its declaration in Alma Ata in 1978, did not change much from the colonial state's suspicion of the traditional midwives. The training of dais and its effectiveness focussed not on enhancing or adding to their knowledge and skills to care for the welfare of the pregnant woman and mother, but on cleanliness and hygiene they were purported not to possess. Further, the dai's role was seen in reducing maternal mortality, the evidence for which was difficult to obtain. They were plainly reduced to obtaining 'cases' for the fulfilment of family planning targets. These half-hearted attempts to train TBAs since the 1960s and 70s were on the wane in the 1990s, as the funding organizations of the reproductive and child health programme discontinued support for training and laid greater stress on skilled birth assistance and institutional deliveries (Sadgopal 2009). This left the traditional midwives lurking within the existing social relations of the village community.

Chinnamalai said her contribution was appreciated in the village but their status as a caste has not altered on that count. Many from the nadar community still treat the barbers as a caste with contempt, although many also respect their medicinal skills. She also charges a minimal amount for the treatment she offers, saying that only if she takes money will her input have effect and that the value of her skill needs to be established. At times her husband jokes, noting the influx of people coming to her for treatment, that she should put up a board outside stating each remedy and its cost.

The possession of simple medical and surgical skills, which are called upon occasionally by people in the village, has only a token acceptance within the caste structure. Despite this, people do come and demand small services. For instance, notes Annathayi, a former midwife who is occasionally called upon for ritual attendance at deaths or births, women would just come by to have a small wound cleaned up so that she would have to drop her own work and attend to them. Thus caste-based services, while not having any social premium in village life as such, are still called for when the upper castes have a need for it.

What remains to be further explored are the articulations of those within the barber community about their knowledge and skills. First, its role in negotiating status within the caste hierarchies of the village community; second, its potential for women within the barber caste to stake a claim for their skills; third, the quest for dignity and respect of those who possess knowledge which is at risk of being obliterated, denied a space within knowledge paradigms, or appropriated by other structures that deny them a space within.

The Women's Movement and Traditional Healers

It is imperative to locate the above discussion in terms of responses of the women's health movement in India. The issues addressed by the women's health movement have been a consequence of the visions of development of the Indian state and an implementation of its agenda of modernity. A child of modernity with its own inherent modernist bias, the women's health movement has been mostly preoccupied with addressing the fallout of state policies, such as the family planning programme launched in the post-independence era and the coercive programmatic means adopted to control the population (Ram 1998; Chawla 2006). Resisting the state's intervention into women's lives in the name of national development and supporting women's rights to their bodies, thus, became for the women's movement the primary goal of women's emancipation. While in pursuit of this struggle against scientific methods that the state adopted through technological interventions such as hazardous contraceptives, vaccines, injectibles and implants, there were also dilemmas and omissions that befell feminists. For us, one of these has been how to resolve or accept the ambivalence that has accompanied the relationship of several feminists to traditional healers, whose existence and practice has always reflected the concern people must have felt for the risky situation of childbirth in pre-modern times. The situation of healers within tradition, culture, and religio-spiritual metaphors and practices amplified this ambivalence on part of most feminists, as these terrains were also entangled with discriminatory and oppressive features directed against women.

Some feminists, however, took to exploring traditions of healing whether through a search for alternatives that tried to chart traditional healers and their use of herbs, as also women-centred methods of self-help to give greater control over their bodies (Shodhini 1997; Sabala and Kranti 1995). Others sought to recover the knowledges of traditional healers, specifically dais,[5] given the neglect that these healers faced not just from the modern medical establishment but the ambivalence displayed by most feminists. In a way these steps reflected the changes in our understanding of medicine, healing and care in the light of the more visible efforts that sought from the state safe, legal, hygienic, yet non-coercive means of reproductive health services, including abortion and delivery. These small but growing efforts also prompt suggestions towards the integration of traditional midwives into state efforts that belatedly acknowledge indigenous medical and healing systems and practices such as ayurveda, yoga, unani, siddha and homeopathy in the Eleventh Plan (Sadgopal 2009). How long this new patronage for indigenous systems will be sustained and succeed in cracking the general negative attitude of the bio-medical system is a question that awaits answers.

In the meanwhile, the more visible feminist understandings and efforts stand uneasily poised with the resistance to newer onslaughts via increased medicalization and commercialization of childbirth witnessed in the debates around assisted reproductive technologies (ARTs), surrogacy, the abortion debates, and so on. In dealing with the issue of health and well-being of women in India, feminists have placed it within the political economy of their daily lives. It is perhaps only feminists among all social analysts who place women and men as equals before enquiring into their overall subordination, thereby uncovering the social trap that systemically reduces women's roles in social production as supplementary labour, consequently compressing what is due to them in the entitlements that independent labour is supposed to get as part of society's and the nation's gross production. The beginnings of this trap evolves from the relegation of women to domestic labour, its invisibility in social reproduction, and the sexual division of labour that it reflects in every sphere where women are employed. This also stamps women with a secondary status, with a lack of self confidence and devoid of autonomy that leads to bargaining power. Ironically, the radical potential that women have lost in this manner is displayed among those women whom society has systematically criminalized, and stigmatized: for instance, women in prostitution, women criminals, women relegated to polluted labour in whom there is a conjunction of the productive (capacity to earn), reproductive (care) and sexual (autonomy to use it) capacities that they utilize for their own benefit (Sangari 1993). This also deeply influences the policy universe that underplays or renders invisible women's paid labour in all the sectors, in addition to their unpaid and reproductive labour in the organized, familial and informal sectors. Given this sketch of the degradation of women's labour, the place of the dai in contemporary discourse is all too evident.

Time-use surveys have tried to capture and highlight the quality of women's productive and reproductive tasks without which significant familial and social systems would breakdown. The women's health movement has significantly engaged with issues around women's reproduction: specifically patriarchal control of women's reproductive labour in her child-bearing capacities, by both the nation and the family. The body of the woman became the marker of the family and the state's need to control its population. Its demographic needs rested on women's personhood. Significantly, while women's maternal role figured prominently in the agenda of the state, their productive contribution as labour to the nation never figures in the accomplishments of the state. Intervening in women's reproduction in order to render her body either capable or incapable of productive labour are contraception, abortion and reproduction with which feminists have consistently engaged, in the health movement for freedom and bodily liberation. The struggle to resist state measures at population control while at the same time having for women an access to safe birth control measures, had several feminists engage with issues of science and technology, body and nature, and the politics of the state and the family (Chayanika et al. 1999). Thus feminists began their engagements with modernity and patriarchy, as well as the conceptual values of modernity: choice, freedom, autonomy. This struggle itself has been all encompassing, resisting the demonic efforts of the state to eliminate poor women in its efforts to control numbers to achieve national goals; combined with the aforementioned ambivalence, it has led to a neglect of traditional healers and healing practices. The women's health movement's neglect of the traditional healers and birth attendants went striding alongside the state's drive for institutional deliveries to reduce and prevent maternal mortalities. While this has been part of most feminists' concerns, other feminists, as noted above, have devoted energies to recover some of these lost resources and capabilities. Acknowledging this relationship as ambivalence, feminists and feminist NGOs are gearing up to new challenges from the state and the market.

Women's Studies scholars have begun engaging with the politics of knowledge-making with systems like science despite their own locations within western modernity. The concern has been to own science, retain scientific temper to grapple with patriarchy and superstition, while also looking critically at the politics of knowledge generation within powerful institutions of science. For feminists this has especially been significant within bio-medicine.

The women's health movement in the last decade of the twentieth century and in the twenty-first century has had to grapple with the role of medical and reproductive technologies, and women's access to them that brought forth the contested questions of female subjectivity and representation for the women's health movement. This is especially the case in issues of sex-selective abortion or surrogacy. Campaigning continues at both a legal and social level in enhancing the conceptions of women's labour into the reproductive. As feminists today we are confronted with newer assertions by women prompted by the market, the state, or the family, which have complicated the notions of labour, specifically for women of the disadvantaged classes. Women migrate, are displaced, end up leaving homes and families, forming new self-sought livelihoods through sexual, domestic and reproductive labour. These new articulations and assertions have made feminists expand their own understandings as well as draw from the past or hitherto unexplored domains in conceptualising labour.[6]

In the run-up to the formulation of the Assisted Reproductive Technologies (Regulation) Bill, 2014, which has carried on to the present, when this volume is going to press, by the Government of India, which is up for public discussion, feminists are yet again engaged in debates of the availability and use of technology for social decisions.[7] State regulation it seems purports to regulate technology but actually ends up shielding the interests of those wielding it. One of the recurring issues that echo in these debates is the recognition or lack thereof of the labour of women and its value in social relations. Thus feminist resources have to keep spreading thin as newer challenges emerge with the medicalization, 'technologization' and commercialization of bodily processes, even as marginalization of existing knowledges continue. In the midst of this, contemporary conditions of degenerate social relations of labour prompt another route to recover the role of the dai or maruthuvacci as a practitioner and healer.

Notes

1 This piece was prepared while researching for the working paper 'Shifts in Women's Work and Ideology in South Tamil Nadu: Labour Process and Gender Relations', prepared during the author's Junior Fellowship (2009–11) at the Nehru Memorial Museum and Library, New Delhi.
2 See Vandana Shiva (1989); Meera Nanda (1996); Gita Chadha (2005); Maithreyi Krishnaraj (1991); Veena Poonacha and Meena Gopal (2004); and Neelam Kumar (2008).
3 Research Centre for Women's Studies (RCWS), SNDT Women's University and Forum Against Oppression of Women (2005, 2006). In June 2005, the Maharashtra government proclaimed a ban on women dancing in dance bars in Mumbai and subsequently in Maharashtra, which became effective on 15 August 2005. The Home Minister and Deputy Chief Minister, R. R. Patil stated on record that these women were corrupting youth who were squandering the hard earned money of their fathers by visiting these bars. Women bar dancers mobilized themselves and formed the Bharatiya Barbala

Traditional Knowledge and Feminist Dilemmas   41

Union, and along with some women's and human rights groups went to court to challenge the ban. The Bombay High Court declared the ban unconstitutional in February 2006, but the state government appealed the verdict in the Supreme Court and sought a stay on the opening of the bars and women reclaiming their livelihoods. The stay was vacated when the Supreme Court in July 2013 upheld the verdict of the Bombay High Court after 7 long years, and declared the ban unconstitutional. The Maharashtra government banned dance bars again in 2014 by an Ordinance, but this too was found unconstitutional by the Supreme Court in October 2015, allowing Mumbai dance bars to reopen.

4 This corroborates what Cecilia Van Hollen (2003) observes as the almost missionary zeal with which medical practitioners campaigned for institutional deliveries.

5 The efforts of health activists and feminists, among others, in initiating projects such as MATKA and JEEVIKA are an important part of this movement (Chawla 2006; Sadgopal 2009).

6 What Sunder Rajan (1993) brings out in her analysis of the subject of sati is that there is a projection of alienation of the subject from pain in western philosophy which is contested by feminist analysis to situate the embodied nature of pain and the need to represent it–that pain constitutes subjectivity, and that the subject of pain is dynamically acting and reacting, rather than being passive and inviting it; for the female subject the body is not alienated in pain. This manner of approaching subjectivity and hence agency will relieve us of the necessity to prove the motivation of women and notions of voluntariness.

7 Note the discussions on surrogacy in the context of Assisted Reproductive Technologies (ARTs) especially articles and studies by Sama (2010) Imrana Qadeer (2009) and Chayanika Shah (2009). The Draft ART Bill 2010 is available for public discussion at url: http://icmr.nic.in/guide/ART%20REGULATION%20Draft%20Bill1.pdf. accessed on 31 August 2015. At the time of going to press, while the fate of the ART Bill is not known, a Surrogacy (Regulations) Bill, 2016, passed by the Union cabinet is pending for discussion in Parliament.

References

Agarwal, Anuja. 2006. 'Family, Migration and Prostitution: The Case of Bedia Community of North India'. In *Migrant Women and Work, Women and Migration in Asia*, edited by Anuja Agarwal. vol 4. New Delhi: SAGE.

Chadha, Gita. 2005. 'Towards an Informed Science Criticism: The Debate on Science in Post-Independence India'. In *Culture and the Making of Identity in Contemporary India*, edited by Kamala Ganesh and Usha Thakkar. New Delhi: SAGE: 246–58.

Chakravarti, Uma. 2003. *Gendering Caste: Through a Feminist Lens*. Kolkata: Stree.

———. 1996. 'Wifehood, Widowhood and Adultery: Female Sexuality, Surveillance and the State in Eighteenth-Century Maharashtra.' In *Social Reform, Sexuality and the State*, edited by Patricia Uberoi. New Delhi: SAGE.

Chawla, Janet, ed. 2006. 'Mapping the Terrain: Birth Voices, Knowledges and Work'. In *Birth and Birthgivers: The Power behind the Shame*. New Delhi: Shakti Books: 11–79.

———. 2002. 'Hawa, Gola, and Mother-in-Law's Big Toe: On Understanding *Dai's* Imagery of the Female Body'. In *The Daughters of Hariti: Childbirth and Female Healers in South and Southeast Asia*, edited by Santi Rozario and Geoffrey Samuel. London: Routledge.

Chayanika, Swatija, and Kamaxi. 1999. *We and Our Fertility: The Politics of Technological Intervention*. Bombay: Comet Media Foundation.

Chowdhry, Prem. 1996. 'Contesting Claims and Counter-Claims: Questions of the Inheritance and Sexuality of Widows in a Colonial State'. In *Social Reform, Sexuality and the State*, edited by Patricia Uberoi. New Delhi: SAGE.

Dreze, Jean, and Sen, Amartya. 1995. India: Economic Development and Social *Opportunity*. Delhi: Oxford University Press.

Geetha, V., and S. V. Rajadurai. 2008 (1998). *Towards a Non-Brahmin Millennium: From Iyothee Thass to Periyar*. Kolkata: Stree.

Gopal, Meena. 2009. 'Caste-based Occupations, Gender and Labour Market Discrimination'. Paper presented at the *Indian Society of Labour Economics 51st Annual Conference*. Patiala: 11–13 December.

Gupte, Manisha. 1987. 'The Social Trap'. In *Search of Our Bodies*, edited by Kamaxi Bhate, Lakshmi Menon and Manisha Gupte. Bombay: Shakti.

The *Hindu*, 2013. Maharashtra firm on dance bar ban. July 23, url: http://www.thehindu.com/news/national/other-states/maharashtra-firm-on-dance-bar-ban/article4942650.ece?ref=relatedNews accessed on 31 August 2015.

Ilaiah, Kancha. 2007. *Turning the Pot, Tilling the Land: Dignity of Labour in Our Times* Delhi: Navayana Publishing.

International Institute for Population Sciences (IIPS) and Macro International. 2007. *National Family Health Survey* (NFHS-3), 2005–06. vols 1 and 2. Mumbai: IIPS.

Krishnaraj, Maithreyi. 1991. *Women and Science: Selected Essays*. Bombay: Himalaya Publishing House.

Kumar, Neelam. 2008. *Women and Science in India: A Reader*. New Delhi: Oxford University Press.

Malhotra, Anshu. 2006. 'Of Dais and Midwives: 'Middle Class Interventions in the Management of Reproductive Health in Colonial Punjab'. In *Reproductive Health in India: History, Politics Controversies*, edited by Sarah Hodges. Hyderabad: Orient Longman: 199–226.

Mencher, Joan. 1978. *Agriculture and Social Structure in Tamil Nadu: Past Origins, Present Transformations and Future Prospects*. New Delhi: Allied Publishers; cited in Ragupathi, K. 2006. *Aadi Maruthuvar: Savarathozhilalaraakkapatta Varalaru (Ancient Doctors/Healers: The History of the Making of Barbers)*, (in Tamil). Pondicherry: Vallinam.

Nanda, Meera. 1996. 'Science Question in Post-Colonial Feminism.' *Economic and Political Weekly* (henceforth *EPW*) 31, 16–17 (April 20): ws2-ws8.

Nihila, Millie. 1999. 'Marginalization of Women Workers: Leather Tanning Industry in Tamil Nadu', *EPW* 34, 16–17 (April 17–24).

O'Hanlon, Rosalind. 1985. *Caste, Conflict and Ideology: Mahatma Jotirao Phule and Low Caste Protest in Nineteenth-Century India*. Cambridge: Cambridge University Press.

Omvedt, Gail. 1994. *Dalits and Democratic Revolution: Dr. Ambedkar and the Dalit Movement in Colonial India*. New Delhi: SAGE.

Poonacha, V., and Meena Gopal. 2004. *Women and Science: An Analysis of Factors Influencing Access and Retention in Scientific Careers*. Mumbai: SNDT Women's University.

Prasad, R.R., and G. Rajanikanth. 1991. *Development of Scheduled Caste Leather Artisans: (Profile, Problems and Prospects)*. New Delhi: Discovery Publishing House.

Qadeer, Imrana. 2009. 'Social and Ethical Basis of Legislation on Surrogacy: Need for Debate.' *Indian Journal of Medical Ethics* (January–March); url: http://ijme.in/issue171.html accessed on 3 April 2011.

Ragupathi, K. 2006. *Aadi Maruthuvar: Savarathozhilalaraakkapatta Varalaru (Ancient Doctors/Healers: The History of the Making of Barbers)*, (in Tamil). Pondicherry: Vallinam.

Ram, Kalpana. 1998. '*Na Shariram Nadhi*, My Body Is Mine: the Urban Women's Health Movement in India and its Negotiation of Modernity', *Women's Studies International Forum* 21, 6: 617–31.

RCWS, SNDT Women's University and Forum Against Oppression of Women (FAOW). 2006. *After the Ban: Women Working in Dance Bars of Mumbai*. Mumbai: SNDT Women's University.

———. 2005. *Background and Working Conditions of Women Working in Dance Bars in Mumbai*. Mumbai: SNDT Women's University.

Rege, Sharmila. 1996. 'The Hegemonic Appropriation of Sexuality: The Case of the Lavani Performers of Maharashtra'. In *Social Reform, Sexuality and the State*, edited by Patricia Uberoi. New Delhi: SAGE.

Rosario, Santi, and Geoffrey, Samuel, eds. 2002. *Daughters of Hariti: Childbirth and Female Healers in South and Southeast Asia*. London: Routledge.

Sabala and Kranti, 1995. *Na Shariram Nadhi: My Body is Mine*, edited by Mira Sadgopal, Bombay.

Sadgopal, Mira. 2009. 'Can Maternity Services Open up to the Indigenous Traditions of Midwifery? *EPW* 44, 6 (April 18): 52–59.

SAMA, 2010. *Constructing Conceptions: the Mapping of Assisted Reproductive Technologies in India*. New Delhi: Sama Resource Group for Women and Health.

Sangari, Kumkum. 1993. 'The "Amenities of Domestic Life" Questions of Labour,' *Social Scientist* 21, 9–11 (September–November).

Shah, Chayanika. 2009. 'Regulate Technology, Not Lives: A Critique of the Draft ART (Regulation) Bill.' *Indian Journal of Medical Ethics* 6, 1 (Jan–March); url: http://ijme.in/issue171.html accessed on 3 April 2011.

Shiva, Vandana. 1989. *Staying Alive: Women, Ecology and Development*. New Delhi: Kali for Women.

Shodhini, 1997. *Touch Me, Touch-Me-Not: Women, Plants and Healing*. New Delhi: Kali for Women.

Siddiqui, Majid H. 2003. 'A Review of Viramma'. In *Gender and Caste, Issues in Contemporary Indian Feminism*, edited by Anupama Rao. New Delhi: Kali for Women: 136–40.

Subramaniam, Vidya. 2010. 'Throwing off the Yoke of Manual Scavenging,' *The Hindu*, Wednesday, 27 October, Delhi edition: 14.

Sunder Rajan, Rajeswari. 1993. 'The Subject of Sati'. In *Interrogating Modernity: Culture and Colonialism in India*, edited by Tejaswini Niranjana. Kolkata: Calcutta: Seagull Books: 291–318.

Swaminathan, Padmini. 1998. 'Failures of Success: Tamil Nadu's Demographic Experience', *Radical Journal of Health* 3: 1 (Jan-March).

Thurston, Edgar. 1909. *Castes and Tribes of Southern India*, vol 2. Madras: Government Press (reprint 1987. New Delhi: Asian Educational Services); cited in Ragupathi, K. 2006. *Aadi Maruthuvar: Savarathozhilalaraakkapatta Varalaru (Ancient Doctors/Healers: The History of the Making of Barbers)*, (in Tamil). Pondicherry: Vallinam.

Van Hollen, Cecilia. 2003. *Birth on the Threshold: Childbirth and Modernity in South India*. Berkeley: University of California Press.

Verderese, M.D., and Turnbull, L.M., eds. 1975. *The TBA in MCH and FP: A Guide to Her Training and Utilisation*. Geneva: World Health Organization.

16

Journeying through an 'Alien Terrain': Feminist Research Methodology and Local Knowledge of Soil Management

Meghana Kelkar

Soil, water, plants and animals form the foundations of agro-ecological knowledge in a given location. But knowledge related to agriculture is multifaceted and complex, spread across a web of agro-ecological and socio-political contexts. Defining 'local' or 'context' is itself problematic as knowledge systems extend from the 'rural household' to the 'global village' with interconnected networks in between. My Ph.D. thesis, 'Local Knowledge and Gender: The Case of Soil Management' (Kelkar 2008) explored the local understandings of soil management, including the local classifications of soils and lands and their valuations within the study locations in two villages of Satara district in western Maharashtra. While recognizing that spatial and temporal parameters are limiting, this cross-disciplinary research concentrated on the local knowledge about soil management and sought to understand the knowledge processes. The local

---

This paper is based on Meghana Kelkar's (unpublished) Ph.D. thesis: 'Local Knowledge and Gender: The Case of Soil Management' (2008), presented to the Tata Institute of Social Sciences, Mumbai. It was extracted and edited by Sumi Krishna, with the author's consent. The thesis was done under the supervision of Professors Chaya Datar (Department of Women's Studies), Nitya Rao and Christine Okali (School of Development Studies, University of East Anglia, UK).

classifications were compared within villages and also with the external schema of soil and land classification– that is, the soil survey reports generated by the district soil survey unit of the agricultural department. Then, the local classifications were linked with the local practices of soil management. Individual practices of soil management were discussed in terms of organizing resources of labour and technology and the socio-political skills that are essential for this pooling of resources. The research included case studies of individual farming households and the micro-level intricacies and actor processes related to soil management, as also the interface between local and external knowledge systems and gendered institutional norms, that impact the valuation of local knowers and knowledge.

Drawing upon my Ph.D. thesis, in this paper, I deal with my own journey as an agricultural scientist doing feminist social science research. In the first section, I discuss the ontological underpinnings and epistemological quests, contrasting them with the mainstream research on soil management. Next, I discuss the methodological justification for 'ethnography', a qualitative research method that has both conventional and feminist forms. Then I sketch some aspects of the process of fieldwork and go on to discuss the multiple techniques used to collect and analyse data. Finally, I reflect upon the subjective dimensions of the research and the ethical dilemmas that I faced during the process.

Questioning the Epistemologies

Epistemology is the study of claims to knowledge. My initial training in natural science did not expose me to the basic frameworks of knowledge creation, the ontological underpinnings and the epistemological quests of a research activity. Thus, this venture of tracing the epistemological roots of my research becomes a journey through an alien terrain. Any attempt to transgress the boundaries of natural sciences and further enter into the world of social enquiry, however, makes this journey an essential and conscious exercise.

In the prevailing classic scientific approach, researchers are presumed to be working in a socio-political vacuum observing strict adherence to notions of neutrality and objectivity. The problem is essentially located in the 'object', the soil, with a marked divorce from the 'subject', the farmer/scientist/extension worker. The object is further reduced into various components (the chemical, physical, biological aspects of soil and so on), and each of these is studied separately, using methods that include experiments and surveys. The results of this research are then gathered, aggregated and assimilated to construct a holistic image, the 'reality'. The knowledge generated is technical and this is used to formulate the research and extension agendas. The research outcomes are considered singularly to develop technologies, with farmers becoming the recipients of technology. Thus, classic scientific research on soils and their management is rooted in the ontological and epistemological underpinnings of natural science such as realism (that assumes an underlying scientific truth) and positivism (that is more concerned with science as prediction).

In the Indian context, the research on soil management is largely limited to the classic scientific approach. This research assumes value-free objectivity and universality, is divorced from the human subject and hence denies that the identity and position taken by a researcher makes any difference to the research outcome. It is assumed that if the complete and unique reality of soils is revealed with the help of scientific methods, then the techniques to manage soils can also be developed in the same way. This technique can be universally applied, and the question of 'who' employs it makes no difference. Thus, in contemporary soil management research, local knowledge is ignored and 'gender' as an analytical category is seldom used. The research primarily addresses the concerns of peer scientists, say, on soil fertility or soil conservation.

Recently, the social constructivist approach has been introduced; this uses a family of participatory approaches and methods, puts the 'farmer first' (Chambers 1983), and focuses on local and scientific synergy in technology design. It would be erroneous to assume that participatory approaches are built on the foundation of a coherent monolithic epistemology. However, this approach does depart from extreme realism and acknowledges that 'truth' can be constructed variously by different actors and institutions, that is to say that reality is socially constructed. Although the ideology of 'participation' has caused an epistemological turbulence in soil management research, this turbulence is neither universal, nor is it toppling down the realist foundation of much soil management research.

The epistemological positioning of my research departs significantly from the mainstream technology-oriented agenda and can be classified as a social enquiry and a feminist project. This is so, not only because my focus of enquiry is 'gender' but more because I locate the epistemological foundations of my research within feminist epistemologies.

Scholars from various backgrounds (many of them being practising scientists themselves) have critiqued both the ontological as well as the epistemological positions of western scientific research (see Keller 1985; Harding 1986; Haraway 1988; Shiva 1988; Marglin and Marglin 1990; Escobar 1995; Demeritt 1996). Amongst these, feminist critiques of science were the earliest to contest the notions of 'objectivity' and 'neutrality'. The location of knowledge in the human mind by philosophies of science was problematised as being disembodied from the knower and the social and historical context in which she/he is located. The fundamental observation was that scientific knowledge is not value-neutral but sociopolitically context-bound and is influenced by the subjectivities of the researcher. This emerging perspective of 'science as a social activity', prone to prefer certain 'knowledge' and 'knowers', was demonstrated by feminist scholars in various ways; from empirical evidence showing the marginalizing and dominating outcome of scientific projects, to the usage of gendered metaphors in scientific language and finally pointing to the epistemological conflicts themselves. Feminists saw science, as a gendered project and critiqued the privileged access to knowledge accorded to scientific communities. They pointed out that these communities are rather exclusive, not only in the sense of excluding people who lack power, but even more so in excluding the perspectives of the less privileged and their ways of knowing (Locher and Prugl 2001).

All feminist critiques of science, however, did not emerge from a consistent singular epistemological position, but rather a variegated mixture of different positions and ontological beliefs. Sandra

Harding summarizes these positions as three simplified strands: *feminist empiricism, feminist standpoint* and *feminist post-modernism* (Harding 1986). This 'labelling and boxing' has been highly contentious on account of over-simplification; however, I find this terminology useful to further the detailing of feminist epistemologies.

Feminist empiricism argues that sexism and androcentrism are social biases correctable by stricter adherence to the existing methodological norms of scientific enquiry (ibid.). Feminist scientists are concerned to conduct their investigations in accordance with feminist principles, to resist androcentric currents of mainstream research communities, and in some cases to organize their intellectual production along theoretical lines identified with feminism (Kohsteldt and Longino 1997). Science and its underlying principles are not totally abandoned. 'Bad science' is identified as a problem and it is further believed that 'good science' can still reveal the true reality of nature.

'Feminist standpoint' as a term was first introduced by Hartsock (1983), drawing from Marxist epistemology. It argues that men's dominating position in social life results in partial as well as perverse understanding, whereas women's subjugated positions provides the possibility of a more complete and less perverse understanding (Harding 1986). This position, like the empiricist arguments, does not totally abandon science; however, it asserts that objectivity never has been and could not be increased by value-neutrality. Standpoint epistemologies thus provide a ground to prioritize the 'situated knowledges' (Haraway 1988) of the marginalized.

Unlike feminist empiricist and standpoint epistemologies, postmodern feminist epistemologies reject the notion of unitary reality and argue that whatever we take as reality can only be a representation or a narration. They exhibit a profound scepticism regarding universalizing claims about the existence, nature and power of reason, progress, science, language and the subject/self (Harding 1986). In doing this they abandon the ideals of science in totality including its tools of analysis. They appeal to cast away preoccupations with 'truth', 'reality' and 'objectivity' and are more interested in the politics of power amongst those who seek knowledge and legitimacy (Haraway 1991). Each of these positions has its own tensions and incoherence; however, I will pursue these

Journeying through an 'Alien Terrain' 51

tensions only in the context of linking feminist epistemologies with constructivist positions as both of these have influenced the epistemological foundations of my own research.

Feminism and Social Constructivism

As already discussed, social constructivist epistemologies believe that there is no reality out there, waiting to be discovered, but is constructed socially by different actors and institutions according to their own perceptions. Social constructivists range from 'weaker' positions that believe in a socially constructed reality distinct from objective facts given by nature to 'stronger' positions that deny the existence of objective reality, arguing that 'truth' is what the powerful believe it to be (see Demeritt 1998).

In the social sciences, the poststructural constructivism inspired by the writings of Bourdieu (1977) and Giddens (1984) rests on the ontology of 'becoming' and argues that agency and structure are co-constituted. The ontology of becoming rejects the existence of a priori social reality outside the human endeavors of social constructions. Further, social structures are argued to emerge from human practice and are constituted by them; neither is conceivable without the other. Norms, rules, institutions or language are taken to be the important media of this structural constitution (Locher and Prugl 2001). 'Human agency' is defined on the basis of the 'knowledgeability' and 'capability' of social actors (Giddens 1984). This agency, it is argued, is shaped by larger frames of meaning and action termed as *habitus* by Bourdieu (see Navarro 2006).

Neither feminism nor social constructivism rest on a single consistent epistemological position, and both display internal deviations and variations. For ease in discussion, however, I will treat them as two positions: 'a feminism' and 'a social constructivism'. Both of these have simultaneous overlapping spheres as well tensions within their positions on ontology and epistemology. Scholars from a constructivist tradition acknowledge feminists to be 'the most enthusiastic proponents of social object constructivism' (Demeritt 1998) by noting that most feminists distinguish sharply between socially constructed reality, 'gender'; distinct from objective facts given by nature, 'sex'. However, some social constructivists dismiss the feminist positions on universalizing and essentializing the uneven structural conditioning of gender relations in the form of patriarchy or capitalism that denies any notion of agency. They also dismiss the positions of feminist ontologies that acknowledge the existence of objective 'reality' (especially in the context of nature). Feminists have also been attacking the social constructivist positions, especially those seeking to argue for a social construction of nature. It is argued that to see science as a purely cultural artifact is to see nature in the same light, which is not acceptable from a feminist perspective, especially for those from an ecofeminist tradition. Ecofeminist epistemologies not only argue for an existence of 'real nature' but for one with its own agency and dynamic. As Mellor (1977) puts it, 'just because hu(man)s have a problem about understanding their environment, it does not mean that it doesn't exist'.

In spite of these tensions, some scholars believe that a dialogue between feminism and constructivism is important because the two approaches add to each other and in combination can yield better theoretical and empirical understandings of the world (Locher and Prugl 2001). They further add that constructivism contributes to feminism a theory of agency, whereas feminism contributes to constructivism an understanding of power as an integral element of processes of construction. They argue that social constructivists lack the tools to explain how gender and power reproduce, how and why certain constructs emerge as more influential than others (ibid.: 113). Feminists argue that power is not a quantifiable material resource, rather a relational social construct and 'gender' a primary way of signifying relationships of power (Scott 1986).

Keller (1989) terms the analysis of social construction of gender and social construction of science as epistemological parallels. She further claims that just as the discussion of gender tends to lean towards either of the two poles of 'biological determinism' or towards 'infinite plasticity', so do the discussions of science exhibit similar polarization, either towards 'objectivism', or towards 'relativism'. She further adds that there has been no intersection between these strands until feminist critiques started converging gender and science. She argues for a stable and clear middle ground, for understanding both of these concepts. She also foresees the difficulty in achieving this kind of a middle ground, first on account of the conceptual difficulty in formulating such a position, and second because of public pressure urging each concept towards one pole or the other (ibid.: 34–35).

Ontological Assumptions and Epistemological Quests

At the outset, I declare my research: a project with an explicit feminist agenda embedded within the goal of seeking knowledge. Thus, I slightly prioritize 'gender' over soil management, making this research less interesting to those who are objectively interested in knowing soils. I do believe in an objective reality of soil, with its own bio-physical composition. My ontological position, however, is that this reality can be socially constructed and interpreted variably by different people. I do not take the scientific data related to soils as ultimate facts; rather put them under the gaze of critical reflection. By discarding the hegemony of scientific constructions, I thus depart from extreme positivist ends. Then, the epistemological concern of my research is the 'subject'; the farmers, the extension workers and the scientists and their subjective interpretations of soils and soil management. My ontological understanding about a 'knowing subject' is influenced by constructionist positions on agency and feminist positions on gender and power.

What I borrow from the constructivist positions is the notion of 'agency', one that rests on the knowledge and capability of an actor. However, feminism contributes an important epistemological dimension to the 'knowledgeability' of an actor, in other words the 'epistemic agency' (Alcoff and Potter 1993). Feminism sees this 'epistemic agency' to be embodied and hence gendered (Moore 1994). With this embodiment of agency the experiential and practical dimensions of knowledge become crucially important as are the propositional dimensions, rejecting the mind/body, mental/manual hierarchies of knowledge (Dalmiya and Alcoff 1993). Further the feminist constructivist ontology believes that the gendered agency that produces knowledge is simultaneously constrained and facilitated by structural forces including those of gender. The gender structure may be articulated in overlapping regimes of division of labour and the relations of power (Connell 1987). In brief, understanding the interplay of gendered agency and gender structure in relation to knowledge production is the epistemological quest of this research.

Acknowledging the positionality of 'me' as a researcher and reflecting on the research journey is also an epistemological act on my part. Feminist emphasis on reflexivity goes beyond a simple assessment of biases and limitations of one's work. It demands a critical evaluation of assumptions, ethical considerations, social contexts of research and most important the open admission of subjectivity and partiality of the researcher's creation of knowledge (Reed and Mitchell 2003). I, therefore, find it essential to acknowledge the influence of my multiple subjectivities on the process of this research: as a woman, as a person trained in agricultural sciences and as one who has worked in the state apparatus of agricultural development. My multi-fold experience has led me to follow an epistemological journey, which is also reflected in my research.

To sum up, the epistemological roots of my research have been fed on mixed soils of feminist and constructivist epistemologies. The situation becomes more complex, considering my training in the positivist sciences. Thus, this means treading on a rather rough and discontinuous path, one that I walked cautiously, consciously and reflexively.

Doing Science Differently

As Fetterman (1989) has rightly declared, the problem must precede the method to avoid the trap of having method in search of a problem. In my research on local knowledge of soil management the research objective, shaped by the theoretical and epistemological underpinnings, warranted an in-depth qualitative research design, 'ethnography' in particular. Ethnography involves the researcher participating, overtly or covertly, in people's daily lives for an extended period of time, observing what happens, listening to what is said and asking questions (Hammersley and Atkinson 1995). According to Baszanger and Dodier (1997), ethnographic studies are carried out to satisfy three simultaneous requirements associated with the study of human activities; the need for an empirical approach, the need to remain open to elements that cannot be codified at the time of study and a concern for grounding the phenomena observed in the field.

The openness and flexibility of ethnography was needed, first, to shed the lenses of the mainstream discipline-bound understandings and, second, to enable the uptake, comparison and analysis of differing perspectives. Ethnography further allowed simultaneous data collection and reflexive analysis and was extremely useful in modifying both the research objectives as well as the outcomes of the research in the process. For example, while in the field, I realized that 'soil management' as a stand-alone concept is incomprehensible from a local perspective. The term 'land care and management' evolved as an alternative, which was much closer to the understandings and practices of local people. In my thesis I have discussed the local cognition of this concept and its expression in terms of various practices. However, the point that I would like to stress is that the ethnographic research design was instrumental in accommodating this change in the research process.

Similarly, the 'Farmer's Field Schools' (FFS) that I was attending as an observer as well as a participant was an excellent opportunity to view and analyse the knowledge interface between the farmers and extension workers. This field realization led me further to follow the actions, interactions and dilemmas of the field workers that ultimately shed light on the struggles over meanings attached to the resource, an important element constituting local knowledge.

The research design also helped me to elucidate the nuances of power and agency within the broader framework of resource management. These subtleties were difficult to explore through constrictive and highly structured quantitative research designs. Even within the open palette of qualitative methods, I found the ethnographic methods to be better suited to allow a situated understanding of the phenomenon, such as the use and management of the land resource, wherein understanding the socio-natural context is of equal importance as the process itself. Thus, the research methodology (theory and epistemology) that I embraced for this research satisfied all the requirements outlined by scholars to justify ethnographic field methods.

Feminism added another dimension, which is not included as a part of conventional ethnography. This dimension was the necessity of continuously and reflexively attending to the significance of gender as a basic feature of all social life and of understanding the social realities of women as actors whom previous sociological research have rendered invisible (Dilorio 1982, cited by Reinharz 1992). As discussed earlier, feminism also distinguished my research from classic ethnography that assumes detached and value-free field research in search of an objective truth. Thus, critical reflexivity was consciously added as an important constituent of the research design, which could then be appropriately termed as 'feminist ethnography'.

Doing Feminist Ethnography

Theoretically, doing this ethnographic research was possible at any location, provided there was land under cultivation and men and women farmers managed it. The choice of the field location, Satara district of western Maharashtra, was based on both personal and academic considerations. Satara lies in the west of the Deccan plateau and comprises three extremely diverse geographic and agro-ecological zones with varied social situations. This diversity has led to tremendous differences in land-holding patterns and agricultural development. There is high male out-migration from the hilly zone. All this presented an interesting possibility of exploring knowledge systems and observing transformations in gender relations in the context of agriculture.

Apart from these academic reasons, personal considerations also influenced the choice. I had worked with the farmers of western Maharashtra, both as a student of agricultural science and as an officer in the Department of Agriculture. This experience had made me aware of the broader agricultural contexts and concerns in these parts of rural Maharashtra. Knowing Marathi was another advantage, facilitating smooth and direct communication with the research participants. Finally, my husband worked in Satara town, during the period of field research, making it convenient to toggle between the responsibilities of research and family.

Scales and Sampling

Sampling in qualitative research stands in instant contrast with the statistically informed sampling procedures common in quantitative research. In qualitative research, the logic and strategy that determines the sampling process are more important than the actual sample size (Patton 2001). I began my fieldwork with a 'big net' approach (Fetterman 1989: 42) casually mixing and mingling with whoever was available, operating thereafter at different scales with differing purposes. As such the sampling procedures were also variable across contexts (i.e., the two research villages, Kurli and Saranji) and across scales (community, household, individual). These variable sampling strategies were based on the classifications proposed by Patton (2001). The Focused Group Discussion (FGD) participants were based on homogeneous sampling (such as the FFS or women's self help group) and convenience sampling (informal group discussions). The case study households in both the villages were strategically selected for maximum variation, aiming to capture and describe the central theme of soil management.

Defining Households and Head of Households

The definition of 'household' is highly contested in feminist, anthropological and economic literature. My research does not adhere to the economic model of unitary households that implicitly cover a nuclear co-resident family, with an altruist (male) head and hides the complexity. Instead, I use a flexible definition of household as a co-residential social unit with a variety of structures and forms across and within societies (Harris 1981; Guyer and Peters 1987). Indeed, the working definition of household evolved in the field. The research makes an interchangeable use of family/household for analytical purposes, with due acknowledgement of their theoretical and practical differentiation as argued by feminist social scientists (Deshmukh 1989; Peterson 1994). The most important feature is the study of individual households with a feminist consciousness, with 'gender' being a critical axiom shaping the intra-household power relations and the interplays of 'co-operation and conflicts' amongst actors (Sen 1990). The emphasis on the households to understand the actor strategies resonates with the view that individual actions of men and women cannot be interpreted separately from the social and residential space that they inhabit (Ellis 2000).

Thus, the major focus while sampling these households was to ensure maximum variation of household compositions (joint/single family), gender of household head (woman-headed/man-headed), land holding and livelihood profiles (agriculture and non-agricultural activities). In Kurli, with its heterogeneous caste composition, caste was another important criterion. The selection of households was gradually spread across the first few months of fieldwork, as interesting diversities emerged. The aim of using a maximum variation sampling strategy was to elucidate variation and common patterns within that variation (Patton 2001). Twenty-three case studies were documented and 19 emerged adequately rich in detail, at the time of analysis. The strategy used to select the village extension workers (VEWs) could best be seen as 'convenience sampling'. The VEWs who were observed and interviewed were those willing to participate and included two women and three men.

The field work extended from May 2005 to May 2006. Initially, I would spend an entire week in a village, take a break of a couple of days to visit my daughter in Pune or my husband in Satara and then shift to the second village for the next week. Later, my visits were based more on judgements rather than a fixed schedule. When in a village, I would spend my entire day, visiting homes, fields, market, and spend the night at one of my case study households.

Techniques of Data Collection

Within the framework of ethnography, I employed multiple qualitative techniques for collecting data such as survey, participant observation, focused and unfocussed group discussions, interviewing, transect walks and document analysis. Participant observation was combined with every other method, to generate reflexive ethnographic interpretations. Triangulation was done across scales, time and space. An ethnographer-researcher cannot adopt a single role throughout her fieldwork. The issue of positioning in the field is also never straightforward and neutral observations are almost impossible. There are complexities of identity, subjectivity and ethics.

I actively participated in daily farming practices, transplanting, weeding, harvesting, and so on. However, I was never 'immersed' completely. My awkward and slow performance usually became a topic of immense entertainment for every observer, but it also paved the way for conversations. My participation in grazing animals on the mountain slopes opened a treasure of information as I could witness copious gossip, observe fields and social interactions, discuss my observations and also ask questions. I realized that this was a hard-earned period of relaxation when men and women grazers were relatively laid back.

I used FGDs mainly to understand the local criteria of classifying-valuing land, concepts and practices of land management and perceptions regarding the agricultural and social change. In both the villages, I used the pre-existing groups—the women's self help group at Kurli and the FFS at Saranji—to initiate discussions. I also organized special FGDs, especially in Kurli where the women's group excluded discussions with men.

None of my interviews was recorded on a tape-recorder as the respondents were suspicious about my motives, spoke blandly or kept asking if the answers were 'right'. The constant use of pen and paper was equally disturbing. I usually jotted down the markers indicating the twists and turns of the interview in my scribbling diary, with verbatim recordings of interesting expressions. After every rich and satisfying interview, I would rush to find a space of solitude and reconstruct on paper the details out of my memory.

Although, this research did not tread into the areas of hermeneutics or semiotics, language still constituted an important data source. Especially in terms of interpreting and decoding informal, 'non-scientific' metaphors and figures of speech in the local language. For instance, when I asked one of my informants to evaluate her soils, she said that they were *halki* (light). When asked to explain, she said: *'You urinate and it is wet, you walk on that urine and it dries! What would you call such land? It is obviously halki!'* (Field notes 2005–06). Clearly, this figure of speech reflected the local mode of communication and also personal criteria of identifying and classifying soils. It contrasted with the more formally articulated modes. As argued by Briggs (1983), it is not necessary for ethnographers to become socio-linguists, but ethnographic understanding should emerge from a broader understanding of the communicative process. Thus, to develop a situated understanding of resource management, it was necessary to understand the local ways of communication inclusive of such in-articulate linguistic connotations.

Data Analysis and Interpretations

My scribbled or elaborately written 'rough' field notes were transferred to a 'fair' field diary, which was more than mere copying because it involved a process of recording the field events, observations and reflexive interpretations to find interesting pointers that could be pursued during the next field session. The observations and interpretations were both about the 'self' as a researcher and the actual research. The household data (the livelihood information, individual interviews, participant observations and so forth) were recorded on paper sheets, forming an individual set for each household. This process of data transfer took a lot of time, energy and effort, and had to be done before the data got 'cold'. Usually I did this in the evenings. However, with frequent night-long 'load shedding' of the electricity supply, the villages started plunging into darkness after 7 pm. The thought of 'losing' my data gave me panic shudders and made me rush towards 'light' in Satara town. Both the 'rough' and 'fair' notes were written primarily in Marathi to preserve the linguistic nuances; field notes were intermittently sprinkled with observations and interpretations in English.

At the end of my field work, I landed amidst a cacophonic pile of papers, note books, flip charts and other documents and yet there was a nagging feeling of something being unfinished and left out. I wanted to fuse this 'left out' part, still 'soft' in my memory, with the 'hard' data that I had gathered during fieldwork. So, I computerized my notes into two coherent themes, the 'village theme' and the 'individual household theme'. The second stage of data analysis was the actual writing of the analytical chapters. This was more difficult as it involved an estrangement from the field and a role transformation from a 'participant field worker' to a 'desk-bound researcher'. This transformation was painful and difficult in terms of creating a sociological text from an ethnographically 'thick description'.

Rejecting standard procedures in qualitative data analysis, I used narrative analysis which was more sensitive to the context and in keeping with the actor-oriented and feminist perspectives that informed the theoretical framework of this research. My analytical framework retains the local structural contexts and also the way people make sense of these contexts, derive meaning out of them and act on these meanings. Yet, the narrative analysis of the ethnographic data does not claim to represent the 'real and pure' accounts of people, untainted by the subjectivities of the researcher.

Ethical Considerations and Protocols

In an interventionist method such as ethnography, the issue of ethical protocols is never straightforward. Acknowledging this intrusive predicament of the research design, I subscribed to the basic ethical principles of field work, that is obtaining informed consent coupled with the right to withdraw, sensitivity to the participant's personal situations, privacy, confidentiality and reciprocity (Punch 1986; Bryman 2004; Mikkelsen 2004).

I obtained the oral consent of all the participants, explicitly sharing information on my personal background and research purpose. The names of both the villages and the research participants have been changed in the thesis to ensure anonymity. Sensitivity is ethically important in field work; at the very start of my field work, both the villages were caught in bouts of heavy rainfall that caused extreme financial loss, in terms of crops and working days. I had to practically stop my work until people recovered from the shock, at least mentally if not financially. In another case, the male head of one of the study households died during my field work. The utter grief of his only daughter and disabled step-mother prohibited my further research inquiries, although it meant loss of a case. But I continued visiting them for social purposes and towards the end of my field work, both of them agreed to participate in interviews.

Many of my participants, especially women, disclosed their intimate private worlds richly coloured with human emotions, violence, love, lust and hatred. They made me acutely aware of the interplay between reason and emotion in the apparently rational arena of land management. However I respected the principle of confidentiality and decided not to purposefully explore this emotional arena further and create distress to my participants. During the field work, I could sense the needs of privacy and discrete 'leave us alone' signalling, especially during torrid fights between family members and other such situations. I remember moving my bed from one home to other, because of the sudden unexpected midnight visit of a migrant husband. Thus the principle of sensitivity was consciously followed to reduce the danger of intrusion.

Reciprocity means much more than attempts to help and expressions of gratitude in the form of minor gifts and *khau*, the practice of visitors presenting children with sweets and savouries. It also means consenting to the research agendas of the local participants, answering endless questions about one's personal life and questions that test one's 'knowledge' by local standards. The ethical dimensions of research get complicated as the subjectivities and the multiple identities of the researcher influence the field activities and create immense dilemmas.

Reflections on the Field Journey: Subjectivity, Identity and Ethics

The feminist consciousness and the constructivist position that has informed my research entail a critical reflexive engagement with the procedural trajectories. The issue of 'positioning' in the field is simultaneously a question of subjectivity as well as one of ethics. Prior to my field work, I had anticipated the predicament of dealing with my simultaneous identities as a 'Ph.D. researcher', an 'agricultural officer' and a 'woman'. However, the actual field work made me realize that there were additional social markers attached to me, such as being 'young', 'married', 'Hindu brahmin', 'mother of a single girl child', making the picture more complex and difficult to 'resolve'. Besides these, there were certain subtleties that aroused equal curiosity, doubt and suspicion: such as 'wearing shoes' (because walking daily through muddy fields made me susceptible to fungal infections), 'not wearing bangles', 'driving a motorcycle', 'speaking Marathi with a *puneri* (Pune) dialect, and 'not drawing *padar* on my head' (using the sari-end as a veil). These subjectivities guided my actions in the field and also created a power field between my informants and me. The social markers that were invisibly but inseparably attached to me could neither be wished off by ignoring them altogether nor be resolved amicably to achieve data objectivity, as suggested in many cookbook-prescriptions on fieldwork. Instead, I chose to 'manage' my various identities to facilitate the best possible field work in terms of data collection, without seriously crossing ethical boundaries. Clearly, this was not an instant job and required continued conscious efforts that lasted till the last day in the field and thereafter. I shall try to elucidate this briefly.

From 'Officer' to 'Student'

Apart from various other methodological criteria, one of the considerations in choosing Satara district for field work was that I had not direct worked there as an 'agricultural officer'. This (I hoped) would eliminate the immediate power-authority influence on both the farmers and the field staff of the agricultural department, the two main participants of the research. Obviously, this did not completely eliminate the 'symbolic' power that gets attached to the position that I hold in the departmental hierarchy. This was obvious in the ways the village extension workers treated me in the early days of my fieldwork. I was 'chaperoned' in and out of the village by one of the extension workers; I was offered a 'chair and table' during the FFS sessions at Saranji while other participants were seated on the mud floor.

This was quite inhibitive in establishing a rapport with the extension workers and the farmers; any friendly attempt of interaction was suspiciously but politely rebuffed. I then resorted to consciously promoting my identity as a 'Ph.D. student' rather than an 'officer'. I changed from starched cotton sarees (considered power-dressing for women officers in Maharashtra) to casual salwar-kameej-dupatta. I insisted on sitting in the 'women's section' during the field school sessions. I made it clear that the purpose of my visits was to 'learn and study' the farming situations in the village and that the study would earn me a degree. I also clarified that I was on 'unpaid long leave' that left me without a direct authority over the decisions and policies of the department. I stayed back without a 'chaperone', long after the field school sessions, visiting households and fields, waiting at the bus stops for long hours and chatting amongst them at public meeting spaces in the village. Both the villages had past experience of being visited by girl students of local colleges for their social work projects. My activities resembled this in various ways and gradually paved the way for my full-fledged entry in the field as a 'woman student'.

In the case of the extension workers, I had to make greater efforts because of the higher preconceived power difference. I consciously took upon the identity of being an ally rather than someone closer to the boss's side. I travelled with them to their work areas, their homes, teashops, offered small help in their work, such as measuring a field, counting the trees, filling the registers, helping in organizing FFSs, and so on. I also took ample care not to be judgmental about their activities and performance, turning a neutral eye to their 'unofficial tactics' of balancing between personal and official worlds. A prolonged period of constant rapport bridged a major portion of preconceived power difference, evident in the ways in which the extension workers began sharing their 'hidden agendas' with me.

Gender and Caste in the Field

Being a 'woman' researcher attached to my other identities resulted in myriad influences on the fieldwork trajectories. Being a married woman with a single girl child earned me lot of sympathy, with elderly women immediately offering to shelter, console and guide me towards a 'thoughtful' second attempt of bearing a boy child. Being a married woman also earned me invitations to Muslim and Hindu cultural events such as *haldi kumkum,* naming ceremonies and marriages (at which the childless or widowed woman was not welcome).

Being a Hindu brahmin woman gave a specific shade, both positive and negative, to my field encounters. Men and women of Saranji associated my 'brahminism' to the profession of 'teaching' and I received several requests to teach English and Mathematics to school-going children. I grabbed this opportunity as it had the potential of opening up further avenues for my in-depth inquiries. In dalit households of Kurli, I was received with a lot of suspicion and discomfort during initial field work. But my literacy was an asset as I could help by writing various applications and decipher difficult documents in English and Marathi. However, with the clear caste-based cleavage in Kurli, I was looked down upon by my upper-caste friends for mingling and eating with the dalits. In Saranji, the otherwise friendly villagers declined to accompany me to the nearby dalit hamlet. Even after spending months in the field, my brahminical ways of cooking, rolling chapattis and my puneri Marathi accent were humoured, mocked and ridiculed. (By the end of my field work I had learnt the non-brahmin way of making chapattis and had largely shed my puneri accent.) The hostility and suspicion reduced significantly as my field work advanced and women introduced me with statements such as, *'Baman hayti pan changlya hayti.'* (She is a brahmin but she is good.) (Field notes, Saranji, 2005–06).

In both Saranji and Kurli, being a 'young' woman was a boon in downplaying the power difference. Especially in Kurli's distinct patriarchal culture, my intimate socializing with women of different castes and active participation in the activities of the women's self-help group led to an instant 'stripping off' of the powerful and authoritative status of an 'officer'. At the most, I was considered at par with the local *anganwadi sevika* (local child-care worker). However this posed another set of problems. In Saranji, I constantly got patronizing warnings from elderly men about the 'dirty' village culture and hints about the impropriety of 'young women of good families' roaming around at odd hours. In Kurli, the gender culture created distinct problems in my field work. My socializing with women bracketed my research as something restricted to the domains of 'women's welfare and child care'. Apart from four elderly Kurli men, the rest avoided any direct contact with me. The moment I entered a house, the man in the house rushed to wear formal dress, exchanged a few words of greetings and left, perhaps with the intention of allowing me a private word with the women. I tried various means of getting closer to men, visiting their homes, requesting them to show me their fields and pertinently sitting in the front hall of the house (a men's space), and addressing them (like the women) with familial names. While these strategies worked to a certain extent, they also raised doubts and suspicion. In the case of one upper-caste household, my constant follow up visits and attempts to socialize with the household head, a young man in his late thirties, created marital discord and tension. My free movement in all spaces in the village and the fields, without drawing a padar on the head, also raised doubts about my credibility as a 'good woman', attracting undue attention from the 'gossip hubs', the village meeting place and bus stop. I was also occasionally subjected to covert comments and lewd remarks. This exposed me firsthand to 'being a vulnerable woman' (read, 'young, without a male caretaker, not abiding to the gender norms of veiling and spatial exclusion') which immensely contributed to the ethnographic analysis of my qualitative data. Nonetheless, these incidents could also compromise my data owing to the exclusion of men.

It is at this stage that I had to resort consciously to establishing an authoritative credibility in the village Kurli. I did this by minor actions such as changing my mode of transport from public rickshaws to a private motorcycle, a conspicuously male symbol of power in the village and undertaking major activities of organizing a series of 'technical' group meetings, with an agenda of discussing the matters of soil management. The venue of these meetings (the village *paar*), completely excluded the participation of women but allowed a direct interaction with men in the village. In these meetings, I explicitly stated my intentions and interest in the issues of land care and management, leading to the surprised expressions from men with regard to my earlier engagement with the women: *'Why did you waste all these days talking to women? What would they know of land care?'* (Field notes 2005–06). This made me realize how the issue of 'soil and land management' got constructed as inherently male, despite the nuances and gender subtleties that I later experienced in my study. Both Kurli and Saranji also had internal political dimensions, based on caste, kinship, class alliances and power patronage.

The above reflections on my fieldwork indicate the 'external' tensions and dilemmas that I faced due to my subjective positioning in the field. Apart from these, there were 'internal' tensions that I experienced on account of my multiple identities that clashed during certain field occasions and also thereafter. How was I supposed to react to the constant mental insults and physical violence inflicted by one of the headmen of my case study households on his wife? Was it ethical to intervene and be honest to my feminist values or was it pragmatic to remain a neutral observer as expected of a classic ethnographer or, worse, support the abuser on grounds of him being an articulate knower and manager of his land? How was I supposed to react to the clearly abusive and corrupt practices of village office holders of various departments? Was it proper to use my official authority and reprimand the practices and the practitioner or was it better to be a 'silent' observer and analyser, being a researcher?

I had overtly introduced myself as a Ph.D. researcher to the extension workers and had also received their oral consent for participation. Despite that I was acutely conscious of the covertness of the observations and interviews that I was recording with them. These were loaded with the contrasting baggage of 'official rule book language' and 'unofficial personal strategies and negotiations'. They did not believe that a seemingly inert topic such as 'soil management' may and can include any of the 'unofficial' material as a research data. Was it then ethical to use this hazy identity and the awkward position between covert and overt? Or was it better to hide the entire set of this data and compromise with the research objectives? These field situations generated constant internal conflicts and a tug-of-war between my identities and ethics. Finally, I started believing in Maurice Punch's emphatic statement (1986: 71):

Field work takes us into a potentially vast range of social settings which can lead to unpredictable consequences for the researcher and researched. The ethical factors associated with the control and regulation of social scientific research are accentuated in participant observation because the field worker has to be interactionally 'deceitful' in order to survive and succeed. Ethical codes fail to solve the situational ethics of the field and threaten to restrict considerably a great deal of research.

The 'culling and refinement' of the field data also created ethical tensions. The wholesome interviews, the extended contacts with the family members of individual households and the rich observations and reflections on their 'life worlds' involved intricate layers of human emotions and rationalities. Grading, boxing and labelling these experiences to generate a finished product and discarding the unnecessary was the most torturous and ethically turbulent exercise that I performed while writing this thesis.

Finally, the research pointed to the need to examine critically the institutional structure of agricultural development, comprising of research, education and extension organizations (see also Kelkar 2009). Most importantly, it is necessary to question the epistemologies and methods of the agricultural sciences in India from a critical social perspective that includes feminist visions. The dogmatic belief in extreme positivism and uncritical realism, the subjective dimensions, the covert norms and value systems hidden under the garb of 'neutrality' and 'objectivity' need to be identified.

References

Alcoff, Linda, and Elizabeth Potter. 1993. 'Introduction: When Feminisms Intersect Epistemology. In *Feminist Epistemologies*, edited by L. Alcoff and E. Potter. London: Routledge: 1–14.

Baszanger, I., and N. Dodier. 1997. 'Ethnography—Relating the Part to the Whole'. In *Qualitative Research: Theory, Method and Practice*, edited by D. Silverman. Thousand Oaks, Ca: SAGE: 8–23.

Bourdieu, P. 1977. *Outline of a Theory of Practice*. Cambridge: Cambridge University Press.

Brady, N. C. 1984. *The Nature and Properties of Soils*. New York: Palgrave Macmillan.

Briggs, Charles L. 1983. 'Question for the Ethnographer: A Critical Examination of the Role of the Interview in Field Work', *Semiotica* 46, 2–4: 233: 62.

Bryman, A. 2004. *Social Research Methods*. Oxford: Oxford University Press.

Chambers, R. 1983. *Rural Development: Putting the Last First*. London: Longman Scientific and Technical.

Connell, Raewyn W. 1987. *Gender and Power: Society, the Person and Sexual Politics*. Cambridge: Polity Press.

Dalmiya, V., and L. Alcoff. 1993. 'Are "Old Wife Tales" Justified?' In *Feminist Epistemologies*, edited by L. Alcoff and E. Potter. London: Routledge: 217–44.

Demeritt, David. 1998. 'Science, Social Constructivism and Nature'; quoted in *Remaking Reality: Nature at the Millennium*, edited by B. Braun and N. Castree. 1998. London, Routledge: 177–97.

———. 1996. 'Social Theory and the Reconstruction of Science and Geography', *Transactions of the Institute of British Geographers* 21, 4: 484–503.

Deshmukh, J. 1989. 'The Household as Theoretical Concept in Economics and as a Unit for Data Collection in the Indian Census'. National workshop on Visibility of Women in Statistics and Indicators: Changing Perspective. Mumbai: S.N.D.T University.

Dilorio, J. 1982. 'Feminist Field Work in a Masculinist Setting: Personal Problems and Methodological Issues'. North Central Sociological Association Meetings, Detroit, Michigan.

Frank. 2000. *Real Livelihoods and Diversity in Developing Countries*. Oxford: Oxford University Press.

Escobar, A. 1995. *Encountering Development*. Princeton, NJ: Princeton University Press.

Fetterman, D. 1989. *Ethnography: Step by Step*. London: SAGE.

Giddens, A. 1984. *The Constitution of Society: An Outline of the Theory of Structuration*. Cambridge: Polity Press.

Guyer, J., and P. Peters. 1987. 'Introduction', *Development and Change* 18: 197–213.

Hammersley, M., and P. Atkinson. 1995. *Ethnography*. London: Routledge.

Haraway, D. 1991. *Simians, Cyborgs and Women. The Reinvention of Nature.* New York: Routledge.
———. 1988. 'Situated Knowledges: The Science Question in Feminism and the Privilege of Partial Perspective', *Feminist Studies* 14: 575–99.
Harding, S. 1986. *The Science Question in Feminism.* Milton Keynes: Open University Press.
Harris, O. 1981. 'Households as Natural Units'. In *Of Marriage and the Market: Women's Subordination Internationally and Its Lessons,* edited by K. Young, C. Wolkowitz and R. McCullagh. London: Routledge: 136–55.
Hartsock, N. C. M. 1983. *Money, Sex and Power: Towards a Feminist Historical Materialism.* New York, London: Longman.
Kelkar, M. 2009. 'Mainstreaming Gender in Agricultural Research and Extension: How Do We Move Beyond Efficiency Arguments?' In *Women's Livelihood Rights: Recasting Citizenship for Development,* edited by Sumi Krishna. New Delhi: SAGE.
———. 2008. 'Local Knowledge and Gender: The Case of Soil Management'. Ph.D. thesis submitted to the Tata Institute of Social Science Research, Mumbai.
———. 2007a. 'Local Knowledge and Natural Resource Management: A Gender Perspective', *Indian Journal of Gender Studies* 14, 2: 295–06.
Keller, E. Fox. 1989. 'The Gender/Science System: Or, Is Sex to Gender as Nature Is to Science?' In *Feminism and Science,* edited by N. Tuana. Bloomington, IN: Indiana University Press.
———. 1985. *Reflections on Gender and Science.* New Haven: Yale University Press.
Kohsteldt, S. G., and H. Longino. 1997. 'The Women, Gender and Science Question: What Do Research on Women in Science and Research on Gender and Science Have to Do with Each Other?' *Osiris* 12 (Women, Gender and Science: New Directions): 3–15.
Locher, B,. and E. Prugl. 2001. 'Feminism and Constructivism: Worlds Apart or Sharing the Middle Ground?' *International Studies Quarterly* 45: 111–29.
Marglin, Apffel F., and Marglin, Stephen A., eds. 1990. *Dominating Knowledge: Development, Culture and Resistance.* Oxford: Clarendon Press.
Mellor, M. 1997. *Feminism and Ecology.* New York: New York University Press.

Mikkelsen, B. 2004. *Methods for Development Work and Research.* Thousand Oaks, Ca: SAGE.
Navarro, Z. 2006. 'In Search of a Cultural Interpretation of Power: The Contribution of Pierre Bourdieu', *IDS Bulletin* 37, 6: 11–22.
Patton, Quinn M. 2001.'Purposeful Sampling. In *Ethnography* II, edited by A. Bryman. Thousand Oaks, CA: SAGE. 106–21.
———. 1990. *Qualitative Evaluation Methods.* London: SAGE.
Peterson, J. 1994. 'Anthropological Approaches to the Household', in *Capturing Complexity: An Interdisciplinary Look at Women, Households and Development,* edited by. R. Borooah et al. New Delhi: SAGE: 87–101
Punch, M. 1986. *The Politics and Ethics of Field Work.* Thousand Oaks, Ca: SAGE.
Reed, M., and B. Mitchell. 2003. 'Gendering Environmental Geography', *Canadian Geographer* 47, 3: 318–37.
Reinharz, S. 1992. *Feminist Methods in Social Research.* New York: Oxford University Press.
Scott, J. W. 1986. 'Gender: A Useful Category of Historical Analysis', *American Historical Review* 91: 1053–75.
Sen, Amartya. 1990. 'Gender and Co-Operative Conflicts'. In *Persistent Inequalities: Women and World Development,* edited by I. Tinker. Oxford, Oxford University Press: 123–49.
Shiva, V. 1988. *Staying Alive: Women, Ecology and Development.* New Delhi: Women Unlimited.

17

'Blue Flower Mentoring':
Interview with Vidita Vaidya

Gita Chadha, Sumi Krishna and Unnati Tripathi

Neuroscientist and behavioural biologist Vidita Vaidya has been with the Department of Biological Sciences at the Tata Institute of Fundamental Research, Mumbai since 2000.[1] *At a relatively young age, she became a Wellcome Trust Overseas Senior Research Fellow and an Associate of the Indian Academy of Sciences. Her biographical essay in Lilavati's Daughters*[2] *begins jocularly, talking about the 'family business'. An only child, she 'grew up on a diet of literature and science', her grandfather and granduncle were 'Gujarati novelists and poets of repute'. Her parents are 'committed medical doctors and scientists' and an uncle is a molecular biologist. She says, 'These two career paths were almost a natural progression in the family business.' Vidita Vaidya works, among other things, on the impact of maternal-separation on infant rodents. She is married, has a daughter; and says that her parents in-law have helped her juggle motherhood and a full-time career in science.*

In 2015 Vidita Vaidya received the Shanti Swarup Bhatnagar Prize for excellence in medical sciences. She is one of the very few women to have received the award since its inception in 1958. Vidita's work on the study of emotion is of central importance to a feminist approach in and to science. Constructions around this kind of work as 'soft' are instructive of the masculinization of science. We also find that Vidita's integrated approach to the life work balance is shared by feminists. Here we engage with her in a conversation on varied issues that concern both women in science and feminists looking at science.

GC, SK, UT: *What attracted you to biology? Your parents, medical doctors, were a great source of inspiration; was there any other reason? It is often argued that women are attracted to biology because they have a special relationship with nature, would you agree?*

VV: When I wrote an essay for *Lilavati's Daughters*, I was joking about joining the family business but it was not completely a joke. I grew up in a research centre campus with scientists pursuing chemistry, drug development, pharmacology—hard core scientists. So I grew up with a lot of research being discussed. I was surrounded by a research culture. That definitely played a part. My parents are both medical scientists interested in biology. When I was six I liked the word 'pituitary'; I thought it was funny. My mum was working on the regression of pituitary tumour then; and I had heard the word so many times that I was running around the house saying pituitary. I grew up with a fair bit of science being discussed on a beautiful campus with snakes and foxes. Among my earliest memories are walks with my parents during which we would discuss animals–why does the owl do this, the fox do that . . . my feel for nature could be traced back to those early experiences, both of being surrounded by a research culture and of being in direct contact with nature.

But no, my parents are not behavioural biologists. My dad is an internal medicine guy, my mother does neuroendocrinology. Yes, the brain is involved but they are not really neuroscientists. They are medical people very interested in clinical research. Let's put it this way–I certainly got very favourable responses when I asked my parents questions because they also found them interesting. The only other conversation that was happening at the table was writing. The big thing in my childhood was my grandfather writing books and reading out from his novels. Story-telling and story-writing was a huge component of my childhood. My grandfather would finish one chapter and when I came home from school, I would lie down, put my feet against the wall and listen to him reading what he had written.

GC, SK, UT: *Was it a privileged life would you say, looking back?*

VV: I took it for granted then, but it was a privileged existence in terms of the exposure and opportunities. It never even occurred to me then. Despite being in a metropolis, it was almost like being stuck in the middle of nowhere. I grew up in the northern suburb of Mumbai called Goregaon. No rickshaw would go from Goregaon railway station to the campus, which was near Aarey milk colony. We routinely had leopards and other animals wandering into the campus. The privileged elite were supposed to be in South Bombay not where we were. But yes we were upper middle academic and professional elites.

Probably, my fascination with behaviour came after I saw a terrific film and talk by the Shanklins, a couple who had studied behaviour in the wild in Africa; they discussed territoriality and aggressive behaviour of the hippopotamus. I must have been ten or eleven. The Wednesday evening movie club on campus usually showed Laurel and Hardy, Charlie Chaplin movies. Everyone would go because it was air conditioned, cosy and a big deal to go to the research side of the campus. I can't remember the other movies but the one on the hippopotamus stuck in my mind. I remember being blown away–how does this animal know how to be aggressive, how does it recognize a threat, why are female hippos so aggressive when they are protecting the young? Then my uncle gave me a book on the brain which I found very fascinating. So yes, growing up on a research campus gave me opportunities...

GC, SK, UT: *Would you say that science is also, in a way, story-telling?*

VV: Yes, very much. Behaviour is magical. Complex behaviour is like reading an incredible story. It was, to me, the most magical, in some ways, of the sciences.

GC, SK, UT: *Can we see a link between what you do and the stories you heard?*

VV: Huge, because I was fascinated by the characters in the stories, who they were, what they did, what drove them to incredible acts of bravery or cowardice? What were the driving forces . . . and a lot of this was also about Indian independence. My grandmother used to tell me about the *prabhat pheries* [dawn walkathons for communal harmony]. In poetry, literature and science, what was magical was behaviour. That interest evolved over time, and in my teens, fifteen to sixteen, I knew I wanted to study the brain. I would have loved it if there had been an opportunity to take psychology along with my courses in life sciences and biochemistry but that was not possible.

GC, SK, UT: *Did you choose biology consciously?*
VV: Very consciously. I knew I was going to do Life Sciences. I did not want to do medicine, my class mates were shocked. It was generally thought that this was the dumbest thing to do–giving up a medical seat. But my dad was very happy with whatever I wanted. My mother said 'whatever you do, make it a conscious choice–perform well. Do not opt for BSc. because you did not get into medicine. My parents would not have had any issue if I had become a writer.

GC, SK, UT: *So your interests were based on nature and story?*
VV: I saw character and personality even in animals. Like dogs in the colony have their own personalities, crankiness, gentility. Even cats have it. So animals carry their own inherent quirks and personality types. Where do all these behavioural traits come from? This fascinated me. And I used to go to my dad's lab so I had seen rats and looked through the microscope. So generally, the idea of being in a lab was very cool. I like breaking things down so there is a reductionist tendency. I remember doing a project on hookworms, collecting them. Underneath the microscope, I liked looking at how the cells functioned. I really enjoyed it. I enjoyed my experiments in school. I liked chemistry and biology. I enjoyed all the science subjects in school.

GC, SK, UT: *You often say that your work is closely linked to psychology. Psychology is also very reductionist.*

VV: It could be. I find the design of some of the original psychology experiments very interesting from an external point of view. I find them very educative, to think about how they can be translated into experiments on animals.

GC, SK, UT: *Are there more women in biology because they have a greater affinity to nature?*

VV: Not particularly. It has never occurred to me that I chose to be a biologist because I had a special relationship to nature as a woman. I find nature fascinating but I can't imagine why any human being would not be fascinated with nature and why women should particularly find it more fascinating than men.

GC, SK, UT: *Arguably because of childbirth and other similar experiences of women?*

VV: I don't think those experiences are necessary to appreciate the mysteries of the natural world. I think they are present every single day and everywhere. These are open to all. I think all children play with ants or whatever is walking around....

GC, SK, UT: *But there are more women in biology—globally and in India—as opposed to physics. Is physics, in the hierarchy of sciences, perceived to be 'harder' (i.e., rigorous, factual, reductionist, dealing with the most elemental truth) as opposed to chemistry and biology?*

VV: It depends on the level at which you ask the biological question. There are biological questions that allow you to use the same reductionist approach more easily, partly because you have taken out the confounding variables as much as you could. There are many more confounding variables in a cell than there would be, say, in a test tube where you can control conditions. Physics just happens to have reached a particular stage in which several insights can be built upon. When we biologists do group comparisons, we use statistics. Sometimes it is not possible to extract a mathematical model out of what you are looking at, and if you could, your mathematical model may not explain the phenomenon. But in cases where mathematical models have been generated, the question can be asked, without a value judgement, whether one level is better than the other. That would be my personal perspective. Phenomenology, reporting an interesting, novel phenomenon, is extremely important. Producing a mathematical model to understand a phenomenon is also important but I would not put one over the other as a better scientific discovery. Two hundred years ago, how well were mathematical approaches applied to all the questions in physics?

GC, SK, UT: *Do you see mathematical models as masculinized i.e., abstract, formal?*

VV: Well, it is a common language that is often used to explain something. It is a language that also helps science to explain a phenomenon.

GC, SK, UT: *Does mathematicization of a discipline put it out of reach, from the general public, from other disciplines?*

VV: There are two ways of thinking about it—explaining something and understanding it better, happens at multiple levels. Mathematics can be a tool to explain a phenomenon. That tool allows you a certain power but it may not always be the most appropriate tool. It depends on what you know about the natural phenomenon. So just because one is not applying mathematics, it does not mean that you are not asking questions about what you are studying. You might be applying a component of mathematics, or you might be applying statistics. There might be phenomena that are understood purely as phenomena without understanding the mechanisms of those phenomena. Simply understanding the phenomena is also interesting. Though mechanistically explaining these is very important, sometimes reporting and understanding a phenomenon per se is also important.

I think the culture over time has begun to realize that it is unhealthy in the long run to put artificial boundaries among physicists, biologists, chemists. At one time science was a hobby for those who were fascinated or who could afford to do it. Later, it was being funded by tax payers, so you had to bring structure, accountability to it. You had to classify, set up departments and programmes; each of these vie for money and funds from a system. We justified biology by saying that we need this funding to solve all medical problems. Yet, very often the biggest medical discoveries have not come from clinical, applied science but from some completely different system with no obvious connection to a medical application. Physics has often been justified by defence needs. The source of the funding affects the practice of these disciplines. Interdisciplinarity can only work if you do not create a hierarchy among the sciences and are not disdainful of your colleagues' efforts, in whatever area, arena. That is one of the reasons why it is least likely to see strong collaborations between the humanities and the sciences because there has been this mutual mistrust. It exists even within sciences, even among biologists.

GC, SK, UT: *Tell us a little about your work in neurobiology.*
VV: I try to understand how circuits in the brain–in the frontal cortex, the hippocampus, the hypothalamus, structures like these–modulate emotional responses and how environment impinges on their normal post-natal development. I am particularly interested in how early life experiences–like change in the interaction between care giver and infant, or a pharmacological agent coming in during early life–shape the development of the nervous system, such that years or months later, the experience/agent leaves a consequence on behaviour and on brain circuits. This is half of my lab. The other half is dedicated to how drugs that influence the mood actually affect the brain, how they eventually bring about changes, and what are the molecular, cellular changes that these drugs create. This is a broad description of my

lab. Fundamentally my interest is in circuits that modulate mood and emotion.

This is very much a trend in the West, partly because of strong clinical reasons and because it also impinges upon psychiatry. Depression, anxiety, schizophrenia and psychiatric disorders are a substantial component of health, so pharma companies are interested. In the West, it has always been an area with a fair bit of focus. Not so much in India partly because few institutes in the country offer masters degrees in neuroscience. Biology in India is largely molecular and cellular biology.

GC, SK, UT: *What could be the reason for that?*

VV: Indian scientific institutions tend to be more hierarchical, partly because the major institutes were built around physics with the money source linked to defence establishments. The power and money structures have driven things differently. Besides in a young, developing country, the priority is to treat patients and to teach the next generation of doctors. Research is not a priority.

Obaid Siddiqi was a neurobiologist, a molecular biologist by training, who was very interested in olfaction. His work was critical for the development of neuroscience in the country. But his joining TIFR to build the discipline was initially opposed. The domination of physics was clear.

GC, SK, UT: *How is your specific study of emotions perceived within the community of neurobiologists?*

VV: My work is perceived as 'soft' science. It is a result of our need to explain things in a reductionist way. Besides, not too many have tried asking these questions. It is relatively easier to explain a single neuron in a dish than look at an entire complex organism and its behaviour. We still don't understand much about emotions. I think emotion seems to cause a reaction of fuzziness. The immediate reaction is Ohh! How can you even quantify this or how can you put a definition to it so why study it. That is worrisome. If that was the case then you would have never started to study most things because in the beginning things were very difficult to define and difficult to put into any kind of a structure. You can break down the entire structure of a musical but you are not supposed to feel the emotions associated with it. I mean a synthesizer producing it versus somebody else generating it, seems to evoke completely different responses. You cannot dismiss that component, it is a huge part.

GC, SK, UT: *And the idea of scientist feeling something themselves while studying emotions-that is a highly unacceptable idea isn't it?*

VV: That's partly because of the strong need to say you are a neutral observer sitting on the outside where you don't get pulled into any of this

GC, SK, UT: *Would you say that this idea is in the basic foundations of science?*

VV: Yes. But it depends on what you are studying: you put it into the context of what you understand–you run the risk of contextualizing. You observe data, and data are data. It is the interpretation of the data where all this comes in. You look at two data sets and that are your data. It stands true whether I interpret it or X interprets it. It is possible that our interpretations may vary slightly in the nuances. I think that depends on the context. Say, two persons with completely different forms of training look at the same data set, their interpretation of how they portray it, convey its importance and relevance will vary. Somebody who is interested in psychiatric disorders may say it is a very interesting biological phenomenon. Somebody else who is interested in learning may say this biological phenomenon relevance is to learning. It is often the window you observe through that makes for the interpretation, the nuances. Data stay the same. But, at the end of the day, when you are telling the story of your data in biology you are contextualizing these to the natural world. Even in physics, it is about the natural world.

GC, SK, UT: *What about the social, within the sciences?*

VV: The strongest examples of this are experiments which were done more than fifty years ago in societies. What were supposedly completely ethical experiments to do then, we now look at them in horror. For instance, The Tuskegee syphilis experiment was a clinical study conducted between 1932 and 1972 by the U.S. Public Health Service to study the natural progression of untreated syphilis in rural African-American men in Alabama. They were told that they were receiving free health care from the U.S. government![3]

GC, SK, UT: *Coming to your work on early-care and separation among animals and extrapolation of your results. What are the problems and how have you addressed these over the years? Have your views changed?*

VV: My PhD work was on models of adult stress, how stress and anti-depressants affect the brain. My post doc was on pathways that influence the brain, trying to understand how some anti-depressants and cell pathways influence brain structure. Later I became interested in vulnerability and resilience. Why do two individuals, even genetically similar individuals with similar environments, respond so differently to the same experience, such that one individual may manifest psychiatric disorder and another may not? War veterans who go through the same level of trauma in war, same experiences, might have very different behaviour trajectories. One might show post-traumatic-stress-disorder and the other might adapt. What makes you vulnerable or protects you from psychiatric disorder? It is clear from studies in psychology and psychiatry that there is a critical period in early life where a lot of emotional circuits are developing and that there is a large impact of environment on the brain. My interest was in asking if early experiences shape vulnerability or resilience.

The 'maternal separation model' is a respected and fairly well-used model that Seymour Levine, Paul Plotsky, Michael Meaney and several scientists worked on; it is not so much a model of disease but of vulnerability. I take away the rodent pup only for three hours. I did not want to do nutritional deprivation on the pups. In this model, a rodent system which is not bi-parental, the pups don't get paternal care, so there is only maternal separation. It is not separation of care givers. It is already a laboratory situation and you are extracting. Unfortunately, the model is called 'maternal separation'. And at that point of time it never occurred to me as a problem. Now I can see the problems because this can be interpreted in so many ways. But it has been routinely called the maternal separation model in the field for years. In terms of actual description it is correct, because it is a dam [female parent] being separated from a pup and the change is in maternal care. Among other rodent species if you take the paternal component away, there can be consequences. If it is done with primates, it is called peer separation or peer isolation. You can ask why the term was used but purely from a factual standpoint, it is maternal separation for a short duration. The problem lies is the anthropomorphization of that data.

When I gave talks on it to audiences who had usually heard about the model, knew that it was only one of many models, understood its strengths and weaknesses, it was not a problem. But it was not so when I talked at the Women in Science meeting in Ahmedabad in 2008. At the end of the talk I was describing what happens to the pups, how they show long lasting changes in behaviour, many of these changes are associated with higher anxiety and molecular changes in the brain. I was not making a deterministic statement that this model is only associated with adverse consequences because our work is now showing that it depends on what you look at, how you look at it, the way you measure it. But while I was describing the anxiety caused in the pups, somebody in the audience, after 40 minutes of listening to me, asked are you essentially saying that all women should stay at home and look after their children? I could see how this could easily be extrapolated. I had not realized that this could be anthropomorphized to this extent. It worried me. I was not saying that at all. I realized that when I am talking to a broader audience, I better caveat it right at the beginning because of the leaps and jumps that are made from the data to a conclusion that might have other implications. I had to explain to the questioner that it is all about care; if you replace the care through another care giver, the anxiety related consequences do not appear.

GC, SK, UT: *Do you correct yourself to an extent or do you dismiss it saying that this is not the right interpretation of my work?*
VV: The problem is that for something to become textbook-level knowledge, it takes the efforts of multiple scientists working, often in teams that are all over the globe, doing their own thing, eventually resulting in some consensus about something. Often there will be opposing pieces of data, arguments, discussions, before these are in the textbook. It takes a long time to develop consensus about what the data are telling you, what you understand about these, how you interpret these and then how much of what you have studied on an animal model is directly applicable or not to humans. The problem is if you simply make statements without the associated caveats, you can extend way beyond what your data actually allow you to state. That is a danger. This is a mutual two-way street; ever since the public started funding science, scientists have to justify their existence to the public because it is the tax payers' money. In the end you are responsible to your tax payers. Very often, a scientist is genuinely driven by curiosity and not necessarily by the application of what may result. Many efforts are application-focussed but very often, in fundamental research, curiosity itself drives the effort. Now, in a changing world where we have to justify why large amounts of money are funnelled into science, there is a need for science to continuously engage with the public to present ongoing work with the necessary caveats so that premature conclusions on an as yet nascent field are not drawn.

This can sometimes be tough because if you present a conclusive statement prematurely one may draw wrong conclusions and run the risk of destroying the mutual trust between the scientific community and the public. The scientific community is only a component of the public. This then results in things like the genetically modified crop crises. The scientific community is trying to protect its right to ask fundamental questions. This is very important. But this cannot be at the cost of isolating yourself so much that you fail to communicate with the public at large. It means that you have to sometimes say that it is incomplete yet and I cannot give a conclusive statement. This space is needed. The gap has widened and has resulted in distrust. Public outreach programmes have to be an active and strong component of scientific institutions. They are a part of scientific responsibility. You can't say, I am doing something that is so esoteric that I do not feel the need to communicate this. That is not justified.

GC, SK, UT: *Let's talk about how gender biases seep into the research design. You are one of the few who is at least open to accepting that there might be such biases.*

VV: Absolutely. I don't think that others are not open to it; any scientist can see the data available out there and ask whether or not an equal number of experiments have been performed among males and females. It would be useful to do such a study. Based on my readings and the years I have worked, even in my field, I can tell you that it is heavily lopsided towards doing studies with male rats. It would be a great exercise to ask factually what this means.

GC, SK, UT: *Very often, there is a naturalization. There is a tendency to study only the male processes or when it comes to the female, it is only the reproductive functions that are studied.*

VV: People who are interested in studying sexual dimorphism, often try to understand the effect of estrogen on the brain and are the ones who have at least in neuroscience exerted the effort to look at both the male and female brains. The question is why has it often been easier to study males? How much is that a consequence of an attempt to take away confounding variables. Species can be a confounding variable, also age. Sex becomes a confounding variable. Often, male becomes a default state and the female is different. Very few studies will be done on aged animals. A lot of the drug studies done in drug bio-availability or kinetics were all done on young adult volunteers. You don't know how the drugs will be metabolized by a child.

I think any scientist should be open to the possibility of addressing this. Till you do the experiment, you don't know if it is trivial. Data will compel a reappraisal of that notion or it will not. It might prove your point of view but it might also disprove it completely. For some things you might find that it does not matter whether you use a male rat or a female rat and for some things it might matter. Till you do that experiment, you don't know.

Where is this discussion going to come from? Either it begins internally, from within the scientific community, and it often does happen. I don't know if that is because there are now more women in science or just more people open to considering such questions, I am not sure. But the general level of openness to dialogue has to stay.

GC, SK, UT: *What has been the role of mentors in your career? Do women make naturally better mentors? Do they have a different style of mentoring?*

VV: No, I certainly don't think that if you are a woman, you'd be a better mentor necessarily. A good mentor is someone who helps you actualize the best that you can possibly be and facilitates that but gets out of way at the right time, does not ever provide you a crutch but is there for you to actualize your best potential. My best mentors have been men. I have some wonderful women mentors as well but my thesis advisor was a man, another man who was almost a second advisor, my thesis committee was four guys. Mentoring is about individuals and what kind of person they are.

It is an unequal relationship; the advisor has a massive responsibility to fulfil towards the student. I don't want to generalize. I have seen examples of both good mentoring in India and some lousy mentoring which I would not even call mentoring. But I have seen examples of both of those in the West as well. I really think that it comes down to the individual. In my personal experience, I have not seen a lopsided proportion of better female mentors *vs.* male mentors.

GC, SK, UT: *Do you think there is a difference that you bring in to your life because you are a woman? As a mentor?*

VV: How much does the consciousness of being a woman matter? How much of it influences what you do? You could contrast it with scientists who are like you in a way but are men. It is difficult for me to distinguish that part of me which is because I am a woman and who I am as an individual. I can tell you which are the things I do differently as a mentor but I wonder if I had been a guy would I have done the same things. It is a very difficult thing to say.

In the twelve years that I have had the lab, I don't think that I have been an equally good mentor to all my students. There are some students whom, I truly believe, I have been able to mentor and others to whom I have been a reasonably good advisor. With some I have not been able to develop a good mentor relationship but have had a reasonable collegial working advisor-student relationship. I have an active engagement with their future career, where they are going, what they need.

GC, SK, UT: *What about the way your lab is run?*

VV: In my lab you are entitled to disagree with me completely and I want a lab where there are healthy disagreements not a culture of 'need to agree with your advisor'. Then there will be no check and balance because if anything I say goes, it becomes a dangerous place. There are individuals who have become my friends, outside of the lab and the professional space. I don't believe that their ability to work with me or to maintain professional relationships is broken down by our ability to have a warm friendship. I don't feel that I have been taken advantage of, as a consequence. As biologists we work in teams, do experiments together and it is a lot of hard work. You are spending your twenties in the lab, you might as well do it in an environment that is collegial and fun. If it becomes drudgery then what is the point.

GC, SK, UT: *Let's talk about the way women scientists are expected to dress.*
VV: I came to TIFR at twenty-nine, wearing a skirt and t-shirt, just like a graduate student from the USA, probably closer in age to the students than the faculty. No one said a thing but you can sense when people want to put you in a pigeonhole. That is pretty standard.

In science, fun and frivolous have become synonymous. It almost appears as if the less fun you have and the more apparent seriousness you bring to it, the better you do. I fundamentally disagree. I am sure you can do everything you do with utter seriousness. I just don't see why you have to take the fun out of it. You are spending a large amount of your time as a scientist doing work that is genuinely driven by curiosity and joy and you don't get paid a large amount to do it. So why should we take away the fun of the discovery process, of being able to laugh at your joys and failures which are a routine in science. What is left if you lose that ability? You can take your work seriously but why must you take yourself so seriously? It is healthy to laugh at yourself and your frailties every now and then.

The risk of enjoying yourself too much is that people will think you are frivolous. It has happened with some of my students. I might have escaped it because I am a vocal, loud and tall person!

Would you call yourself a feminist or a feminist scientist?
I would call myself a feminist but not a feminist scientist. What is a feminist scientist? Is it the same then if you ask me are you secular or are you a secular scientist? I am asking what is the value of the adjective to the science?

Being a feminist might influence the way I am as a person? Yes, absolutely. Influence the way I do science and the practice of my science? No. In the way my being secular will inform my behaviour as the head of a lab, in much the same way my being a feminist will inform my behaviour in the lab.

Let's put it this way, if I was not a feminist, may be it would not have bothered me that the jump from my data to the interpretation was instantaneously the one that all moms should stay at home and never go to work. Maybe I would not have been so aghast. When I look at my data set and ask, is this the interpretation–my being a feminist may not directly impact that. But I would worry about interpretations of the data, how these are converted into something relevant to human kind. Because I still have not understood how it changes the way you look at data. (Here I am saying that I am not sure how it changes data analysis but yes the nuances in the way the data are conveyed that it certainly might.)

GC, SK, UT: *Women in all professions face discrimination at different levels and in varied measures, do women scientists have some unique problems?*

VV: If you have not been at the receiving end, that should not make you blind to its existence. The worst and probably the most harmful thing is putting limitations on ambition or on dreaming–before you even start you curtail the possibilities. This happens in an insidious fashion through subliminal messages that you may be never able to achieve something. This happens at multiple levels, during admissions, in student-advisor relationships and in the kind of projects assigned to students.

GC, SK, UT: *That women can't do a certain kind of project?*

VV: There was one single instance relating to a post-doc applicant, a bright, capable young woman. As we didn't have housing then on the TIFR campus, we suggested that she could be put up at off-campus housing. But we were told that that, it would be unsafe for her to travel late and maybe we should wait till housing opened up at the Colaba campus. Her appointment would have to wait. So she didn't come, as the appointment was deferred. So, this is [what is considered] gender sensitivity, couched in the worst possible manner, very paternalistic! That we should be sensitive to our women, our girls should be on campus. Would you have worried about a male post-doc in this way? We have first made her this weaker individual/sex, who needs to be protected. Thankfully, this rule was overturned later and such an instance did not happen again.

GC, SK, UT: *There is also direct harassment.*

VV: Yes, but direct harassment issues are separate. There is a wide spectrum. This was an example of the more insidious, invisible statements which can sometimes be very harmful.

GC, SK, UT: *What about the instances of direct harassment; instances of violence?*

VV: I have not witnessed them directly. The insidious examples are many more. In institutions where there are good working checks and balances, the most dramatic, glaring instances have not been as many. They may exist in their severity but their number I think has gone down. And I would not diminish the importance of the insidious instances, far more subtle but still corrosive.

GC, SK, UT: *What do you think of the Women's Cells (WC), is it a good state initiative?*

VV: The WC is a very good initiative and I think we should have it in every institute. It provides for a certain protection, a certain awareness and the pro active women cells may even actually allow the possibility of a healthier working environment. You can have a women's cell which works only as a complaint agency–receiving complaints and redressing grievances. An active women's cell would anticipate issues before they occur, and bring to the table things which should be up for discussion.

GC, SK, UT: *Why are there fewer women in science?*

VV: Science is not seen as fun, it is presented as a 'serious pursuit for serious-minded individuals'. And this applies to both boys and girls. Hence globally, we have a problem with the number of people going into science. Another factor here is that girls are continuously being made to feel that they are not going to be good at science. This is what I meant earlier, curtailing the horizon of possibility. It is not that you are actively deterred but you are not actively encouraged either. I don't know how many times an Indian woman scientist is discussed as someone young people can even remotely relate to! The problem gets deeper when we realize that women scientists themselves can intimidate young women in science. One of my mentors, an immensely successful scientist was one of the few women scientists at Yale who had a healthy emotional life. Whereas most of the other successful female scientists I saw at Yale had a pretty touchy personal situation. All this is a consequence, a fallout. You can understand how much pressure there is in being a woman scientist. The ones who have hardened are the ones who have survived; the system has finished off many of the others.

As an example of such experiences being common across cultures and contexts, Vidita Vaidya shared emails she had exchanged recently with her mentor. Vidita shared her essay in the book Lilavati's Daughters *with her mentor. The essay speaks of Vidita's journey into science in general and neuroscience in particular. On reading the essay Vidita's mentor responds by saying that she now has a daughter who is pursuing a career in neuroscience. The exchange of mails interestingly alludes to a chain of shared passions for science between mothers and daughters. Further the conversation revolves around how the sub culture of science is dominated by 'bullies' who inhibit and intimidate women in science while setting up a 'hardened' approach to the doing of science devoid of a wholesome connection between nature, emotion and science. Vidita's mentor refers to how she is tending to the blue flower that had bloomed in her backyard, rather than attending to the demands of her academic*

'Blue Flower Mentoring'

career. Responding to Vidita's appreciation of how she was mentored several years ago as a young graduate students, her mentor says:

To prove I haven't changed, please see photos attached of our backyard, which I took this morning when I should have been writing lectures for the med students (but it will rain tomorrow, so time for it then). I planted the *Chionadoxa* bulbs last fall, so they are my gift in this early spring

Now here is something really terrible to say. I think weak, aggressive men…have always created a world with rules that say that being an asshole is a sign of success and intelligence, not because it really is, but because that is the easy, selfish way to succeed, and thus they are rewarded for their own bad behavior. They are bullies, and there is no adult in the room to put them in the corner. (Economists have done a similar thing by defining "Greed" as the "Rational Solution".) They want to dismiss me for blue flower mentoring, and hate me even more if I have a paper in *Cell*. There is no way to win other than to know that you already have won.

VV: I love the term 'blue flower mentoring'. Around me I see so many women who think they need to be a degree worse than the men to be taken seriously. It wipes out the joy of enjoying the blue flowers, makes for an arid existence in science. I wish there were more blue flower mentors, it would make the practice of doing science much more joyful…

Women scientists have to stand on their heads to be taken seriously. Yale is an extremely competitive environment, it chews people up and spits them out and women just learn to be more tough.

GC, SK, UT: *Your mentor was an exception among the other mentors. How did it shape you as a scientist and as a mentor now?*
VV: It is not that essentially women make better mentors–I had a very good male mentor. It showed me what I wanted to be as a mentor, what I did not want to be. Here we have a very strange amalgamation of *guru shishya parampara* plus power

and authority, a hierarchical system where nepotism also flourishes. We call seniors 'Sir'. For an Indian, what does this Sir mean? Some thought has to be given to India-sensitive practices. What fits what does not. India has to evolve its own scientific practices which are culturally sensitive to our country. Some thought has also to be given to the management and administration aspects. It is not a system which exists in a vacuum where science is universal, there are culturally sensitive nuances. We can't copy, cut and paste systems.

Notes

1 This concept was introduced by Friedrich von Hardenberg (1772–1801) as a symbol of creativity and the search for a higher ideal.
2 Rohini Godbole and Ram Ramaswamy, *Lilavati's Daughters: The Women Scientists of India* (Delhi: Indian Academy of Science, 2008).
3 http://www.cdc.gov/tuskegee/timeline.htm

18

The Healing Touch: Dr. V. Shanta's Journeys in Cancer Treatment and Care

Kamala Ganesh

Everything begins with a personal story.

My first direct contact with Dr. Shanta, at the Cancer Institute (Women's Indian Association) at Adyar, Chennai, was in 2005 when I was detected with breast cancer. I went to the Institute for treatment, not only because my cousin worked there as an oncologist, and I had close family members in Chennai but also because of the many things I had heard about the Institute. During the next eight months, the Institute grew on me, warts and all. The place had a distinct feel to it, quite unlike Apollo Hospital on the one hand and Perambur Railway Hospital on the other, both of which were part of our family health circuit. My first impression was that the Institute was oriented to treating patients from poorer strata; the medical staff, although overworked, were competent in cutting-edge technology and had a sympathetic manner with patients; the general upkeep was about average and care facilities simple though adequate. There were queues and considerable waiting time, but all patients got started on the process of investigation the day they came. Relatives accompanying patients from rural Tamil Nadu and Andhra would be squatting in groups in

---

I would like to thank Gita Chadha, S.G. Ramanan, S. Krishnamurthi, Sarojini Varadappan, Srivatsa Marthi, Sumi Krishna, T.S. Shankar, V. Shanta, Veena Poonacha and Vidya Shankar for enabling the writing of this essay with moral and material support.

the spacious grounds, waiting for instructions on the next step. Despite the crowded atmosphere, one did not get a sense of apathy or callousness. The para-medical personnel were not quite on par with the medical staff, but still things seemed to be working. There was an income-based differential payment system. I quote from an article I wrote in *Tehelka* (Ganesh 2005a) 'As I went through my own treatment, the institute's underlying approach slowly unfolded: priority to saving life at all costs, cut out the frills, advanced technology for core treatment alone, no differentiation between different classes of patients in medical treatment. While cutting-edge technology may be routinely deployed for radiation, total blood count may still be gauged by the relatively old fashioned method of a simple finger prick and manual counting in the 'Neubeaur Chamber'. While better accommodation facilities are offered to those who pay more, the pre-operation/post-operation wards are common to all. . . .'

The overall feel was neither statist nor corporatist, but something else which at that time I was not quite able to decipher. I learnt that it was a voluntary, charitable trust, but it did not have a 'welfarist' feel either. The 'chairman' Dr. V. Shanta had just won the Magsaysay award, and reports and photographs were all over the media. She was my surgeon, and her clarity in diagnosis, calm and concerned manner and firm yet gentle touch infused serenity and confidence in me. When I found out later about her age, I was amazed. 'Even if you get a knock on the head, let it come from a bejewelled hand' goes a Tamil proverb. I remember murmuring to her on the operation table 'Doctor, I am glad that you are the one doing this surgery'.

As a genre of writing, the biography of pioneering institution-builders like Shanta runs the risk of becoming so overwhelmed by individual achievement as to miss the salience of context, limits of achievable success and thorny questions of sustainability and replication of the ethos. Moreover, an approach that focuses on success stories can lead to complacency that 'women have made it'; structural deprivation and injustice for the majority gets eclipsed in the dazzling achievements of a few.

While early feminist historians tried to redress the neglect of women in formal history by retrieving the contributions of 'great women' of the past, this approach has since been subject to sharp scrutiny. The biographies of 'women worthies', points out Jill Matthews, generally portray them as exemplars of masculine standards of womanhood, or as exceptional women who were almost as good as men; it becomes in effect, the history of men disguising itself in women's skirts. Such expressions of attitudes, values and beliefs constituted merely the history of what men thought about women (Matthews 1986: 147–48). Descriptive accounts of women saints' lives, editions of women's writings, and biographies of queens and heroines may contain useful information, but usually avoid difficult questions about the sexual dynamics of power within their societies (Bennett 1989: 254). However, such invocations can also be used as a feminist tool, to celebrate women's past accomplishments and to rebut the accusations of misogynists that women have contributed nothing (ibid.: 251). The key is to contextualize individual lives within a larger socio-political frame that addresses gender inequality as a structure and a system. Life-stories of pioneering women can provide us with insights on the nature of patriarchy and on women's struggles, conscious or unconscious, to make a place within it, and make a difference.

And, yes, of course, there is magic in the single story. It can hook the imagination, kindle emotion and so brings home a powerful message. Some single stories can rise above narrow specificities and verily be read as a comment on a whole world and a whole time: a woman doctor in the first decades after Indian independence in the daunting arena of cancer treatment. An institution built brick-by-brick with an amalgam of Gandhian austerity and Nehruvian science–the former's focus on outreach to the poor–'Daridranarayan'–combined with the latter's emphasis on harnessing technology for national development.

The Cancer Institute and Dr. V. Shanta are synonymous with the early phase of oncology in India, but Shanta's journey is intertwined with that of at least two other individuals: Dr. Muthulakshmi Reddi who established the Institute amidst formidable obstacles and Muthulakshmi's son Dr. S. Krishnamurthi who steered it into a powerful model for the new nation.[1] The Institute is also tied by an organizational umbilical cord to the Women's Indian Association (WIA), which later was a branch of the All India Women's Conference. WIA was set up in 1917 at Madras by three Irish suffragettes, Annie Besant, Margaret Cousins and Dorothy Jinarajadasa, who were also members of the Theosophical Society at Madras. It partook of the major reform and nationalist concerns of the time and played a leading role in getting women the right to vote (Basu and Ray 2003: 19; 69). In this way, the Institute's origins are part of the legacy of the early Indian women's movement, although it is not geared specifically to women's health. Shanta herself comes from an exceptional lineage of scientists. In her biography the histories of social reform and freedom, nation-building and institution-building, science and progress, excellence and outreach are intermingled: issues and discourses which were implicated in the successes and failures of the newly independent India. This contextualization, however, is only by way of nuancing what ultimately is an essay about Shanta's own reflections on her life and work, seen through my own locational lens as her patient and as a researcher who undertook to archive the Cancer Institute for the Dr. Avabai Wadia Archives for Women.[2]

In a significant early contribution to the question of feminist method, methodology and epistemology, Sandra Harding (1987: 1–13) argues against the idea of a distinctive feminist method of research. Preoccupation with method mystifies the most interesting aspects of feminist research processes. It would be more fruitful to identify the features of the most powerful feminist research, which according to her, include recognizing women's experiences as a theoretical and empirical resource, designing research *for* women and placing the researcher herself on the same critical plane of scrutiny as the subject matter, that is, being reflexive. One would qualify further that this is not an exhaustive list, nor is it exclusive to feminist research. It is not only about adding women to a given subject, or necessarily about women. When looking at the female protagonist of the Cancer Institute through a feminist lens, how are the approach and method different from what any sensitive biographer would attempt? It is true that of the three foundational figures of the Institute, two are women, and one of these was a campaigner for women's rights. The Institute itself was established under the organizational rubric of a reputed women's organization, the WIA, which was involved hands-on in the initial stages.

But I would go further than this. The character and thrust of the Institute become comprehensible only through the personal lives of its three protagonists and their inter-relations. The micro worlds of family and friendship and of ideology and emotion fueled the dynamism of the macro arena of advanced oncology. And behind the technological and organizational strengths of the institution lies an ethic of care that has evolved through the personal concerns of the two women: Dr. Muthulaskhmi Reddi and Dr. Shanta. Of the many facets of Shanta as an oncologist and institution builder, I have focused only on these two as being the most pertinent from a Gender-Science perspective.

The essay starts with an account of the Cancer Institute and its founder Dr. Muthulakshmi Reddi. Although deserving of more detailed treatment, it is necessarily brief, as a backdrop to Shanta's own life and work. The major part of the essay is based on my extensive conversations with Shanta, interwoven with events and turning points in her life, available as published information as well as through conversations with her relatives and colleagues.

The Cancer Institute (WIA)

The Cancer Institute, founded in 1954 under the aegis of the WIA by Dr. Muthulakshmi Reddi, was the first specialized hospital for cancer in South India at a time when there was a complete lack of public awareness about cancer as an illness curable with specialized treatment. Indian cancer research and treatment did not command a high opinion in western medical circles. There was little cooperation among the few specialized institutions that existed in India. From the early years, Krishnamurthi and Shanta had to contend with political, governmental, bureaucratic and financial hurdles and popular apathy. Yet, the Institute established itself rapidly as a reputed institution with an emphasis on outreach to the poor and rural strata of society.[3] The Institute was among the first in India to develop a multidisciplinary approach, giving equal place to medical, surgical and radiation oncology. It was the first centre to start a department of nuclear medical oncology in 1957, and the first to introduce radioactive isotopes in diagnosis and treatment. During the 1950–70s, the Institute was ahead of other hospitals in its speed and resourcefulness in acquiring and deploying sophisticated technology. It pioneered radiation treatment, acquiring the first Cobalt 60 unit in South and Southeast Asia in 1957, the first Linear Accelerator in India in 1976, and the first Indian-built Therapy Stimulator in 1969. Contributing their human skills, Krishnamurthi and Shanta also brought in cutting-edge technology by harnessing international philanthropy. They also forged international scientific collaborations which helped them with treatment protocols, provided laboratory support, and funded training programmes for Institute doctors in the United States and elsewhere.

The Institute pioneered specialized academic courses, from 1960 onwards, first through Madras University and later by establishing the Dr. Muthulakshmi College of Oncological Sciences, attached to MGR University, the first such in India. It has produced more than 160 postgraduates in oncology, which is more than the all-India total of degree holders in oncology.

The Institute has contributed to mapping the course, spread and control of the disease through a Hospital Cancer Registry. Despite minimal staff, Shanta developed a protocol for maintaining detailed and meticulous patient records right from the start, with a formidable follow-up system to ensure that patients come for further check-ups. On occasion, the Institute has been known to send its personnel to track a recalcitrant patient and escort him physically for follow up! In a largely illiterate population, this has significantly reduced fatalities. The repository of epidemiological data over fifty years has contributed to a rich institutional memory. The case records system has also enabled patient treatment by a team without relying on the same doctor for every visit. The Indian Council of Medical Research initiated the network of Cancer Registries in 1981, locating the Madras Metropolitan Tumour Registry in the Institute. In 1974, it was selected by the Government of India as the Southern Regional Centre for Cancer Treatment and Research.

The Institute has devoted considerable energy to outreach: prevention and containment of cancer through education, propaganda, field surveys and screenings. Patients and the community at large were benefited by its persuading the Railway Ministry in 1956 to grant concessional train travel for cancer patients. In 1971, it successfully petitioned the Union Finance Ministry to declare anti-cancer drugs as life-saving and allow import of chemotherapy drugs without licence and customs duty.

Starting from a thatched shed with 12 beds, the Institute has now two campuses, 460 beds, and patients from all over the country. It is a recognized centre for treatment, teaching and research. The Institute is like a brand, earning instant recognition and smooth entry for its alumni in professional placements nationally and globally. Its distinct approach has also spread through its alumni. (For a detailed chronology and report see 'Cancer Institute' 2004, on the occasion of its golden jubilee.)

Shanta says she is proud that the Institute has stood its ground with its home-grown approach. She says visitors are surprised when they hear she has trained mainly in Chennai: *'They give value only if you have trained abroad. Yes, Dr. Krishnamurthi and I have taken a stand on this. They say we are mad people!'*[4] She distinguishes the Institute from private hospitals which neither engage in academic activities, nor have research and publications. A hospital, she says, cannot just treat patients if it wants to be scientific, progressive: *'Treatment, research, teaching, preventive oncology, palliative care; we are an integrated institution. We are different, we want to be different.'*

Dr. Muthulakshmi Reddi

The Cancer Institute's unique place among medical institutions in India is the outcome of the missionary zeal and relentless struggle of its founder Dr. Muthulakshmi Reddi. Her personality, social concerns and political engagements shaped its values and ethos. Her approach deeply influenced her son and successor Dr. Krishnamurthi and his colleague Dr. Shanta who steered its development.

Muthulakshmi was one of India's first women doctors, graduating in medicine in 1912 from Madras Medical College. A well-known gynaecologist and obstetrician, she was also an ardent social reformer and nationalist who crusaded for legislation to protect women and children. She was elected Deputy President of the Madras Legislative Council in 1927, and thus became the first woman in the world to preside over a legislative body.[5] She started the Avvai Home for orphaned girls in 1931. She was the force behind the campaign for the abolition of the devadasi system in Madras Presidency, from 1939, when the Bill was introduced in the Madras Legislative Council, till 1948 when the law was enacted. Her mother was a devadasi and Muthulakshmi's personal anger at the injustices and indignities of that identity gave an edge to her ferocious campaign. When Congress leader Satyamurthi argued in the Legislative Council against abolition on the grounds that it was ancient religious custom and that devadasis were the custodians of the traditional arts, she famously retorted that if such a caste was indeed necessary and since the devadasis had done it for so long, why did the brahmin women not take over from then on? (*Stri Dharma* 1928: 48).

Muthulakshmi's own autobiography and writings on her by others vividly describe her multipronged engagement with social reform, politics and public affairs of her time. Her reformist campaigns, lauded by many, have also been criticized by some for being condescending towards devadasis, othering them from a Hindu patriarchal perspective, stereotyping them as immoral and unchaste and denying them subjecthood and agency (Anandhi 1991: 740–41). Yet Muthulakshmi's sincerity and dedication were never in doubt.

Muthulakshmi was a founding member of the WIA, its Secretary for many years and edited its journal *Stri Dharma*. She later became its Life President. Muthulakshmi had lost her sister to improperly diagnosed cancer in 1923. She had nursed her through her last painful days. Amidst her grief, she vowed to establish a specialized hospital for treatment of cancer. She was inspired by the emerging advances in cancer treatment in the West and in 1925 spent a year at Royal Marsden Hospital, London, to specialize in the subject. She got the WIA involved in her mission. It was an unusual issue for a women's organization to take up but the sheer force and dynamism of her personality made this a major activity of WIA for many years, says Sarojini Varadappan, a renowned social worker who was one of the Secretaries of WIA at that time and worked closely with Muthulakshmi. Sarojini (who was later associated with the All India Women's Conference, AIWC), was in her nineties when I interviewed her, and she continued to be President of WIA, till her death in 2013. She remembered Muthulakshmi as an impressive even formidable personality. Immaculately dressed in heavy Kanjeevaram saris, pinned up with a brooch, and wearing shining diamond earrings, Muthulakshmi had a flourishing practice, delivering the babies of all the rich Mylapore professionals.

Initially the WIA campaign was to spread awareness of the importance of early detection and treatment. Muthulakshmi wrote in academic journals and copiously in popular media, and spoke untiringly in various fora for her cause. In 1936, Muthulakshmi mobilized several women's associations in Madras and presented a joint memorandum to the King George Memorial Fund Committee urging that the funds collected should be utilized for the establishment of a cancer hospital.[6] Throughout the following decade, she continued to push the government on the matter. In 1949, in the face of governmental apathy, WIA started a fund raising campaign–Cancer Relief Fund (CRF)–under the presidentship of the Maharani of Bhavnagar, wife of the then Governor, with Sarojini as Secretary and Muthulakshmi as Treasurer. There were exhibitions, pamphlets, posters. Sarojini recounts that initially the (male) doctors were all either against the WIA campaign or simply indifferent. They used to say, why are you wasting your time? Cancer was considered to have no cure, a *karma vyadhi* (fated disease). They used to pass uncharitable remarks against Muthulakshmi.

Eventually, after campaigning for more than two decades, Muthulakshmi managed to get the government to donate a piece of land at Adyar. Prime Minister Jawaharlal Nehru laid the foundation stone in 1952. Sarojini Varadappan, who gave the vote of thanks, narrated to me how Muthulakshmi used all her channels at Delhi to bypass the Madras Chief Minister and make it happen. Sarojini recollected that Nehru in his speech expressed special pleasure that such a worthy activity was being initiated by women. The Institute was inaugurated by C. D. Deshmukh, Union Health Minister in 1954.

Muthulakshmi sent her doctor son, Krishnamurthi, to the USA in 1947 for two years to specialize in cancer treatment. He narrates (The Cancer Institute 2004: 22) how initially, having tasted the comforts of working in the USA, he had no enthusiasm or inclination to become a heroic pioneer, triumphing over ordeals. But eventually, she convinced him by downplaying the argument about service to humanity and prevailing upon him as a son to support his mother's dreams. *'I was very attached to my mother. Then and there we vowed that, come what may, we would see the struggle through together.'* Old associates recall that Krishnamurthi and Muthulakshmi had periods of stormy interaction, but he was devoted to her.

Later, after the Institute was established, the CRF was dissolved. A committee was formed for the Institute, and for a while, WIA office bearers were on it. *'But since it was a medical institution, we all gradually moved out and gave way to specialists'*, said Sarojini. The Institute is still officially known as Cancer Institute (WIA) as a permanent tribute to its lineage.

Muthulakshmi was involved with the Cancer Institute till she died in 1968. During the last decade of her life, with failing health, it was her son Krishnamurthi who was at the helm. He was associated with the Institute right from its inception, being the Secretary of the first governing board. He became Scientific Director in 1956 and Director in 1959. In 1980, he became Chairman and in 1997, he stepped down from direct involvement and became 'Advisor' and remained in the campus till his death in 2010.

The character of the Institute as a voluntary charitable organization has been responsible for some of its difficulties in raising funds, tackling governmental bureaucracy, apathy and endemic corruption. In recent decades, the ideals of simplicity and selfless service have been under attack as being impractical and simplistic in a climate of aggressive corporatization of health care; and discussions that the Institute has not created an effective second rung of leadership have taken place. Of late, there have been severe internal rumblings from the doctors about low pay, long working hours, autocratic leadership and lack of autonomy in functioning. The Institute is grappling with serious questions of relevance, viability, survival and restructuring in a radically transformed contemporary scenario of medical and health care.

Muthulakshmi was the quintessential activist and her contribution lay in establishing the institution in the face of adversities. Her campaigns from the late 1920s onwards form a fascinating tale of passion, persistence, resourcefulness and ingenuity in leveraging her professional and political clout to achieve her objectives. However, in the rapidly advancing field of oncology, it is Shanta with her training and single-minded pursuit of medical excellence and organizational efficiency who can wear the mantle of being a woman pioneer in cancer treatment and care in India.

Dr. S. Krishnamurthi

It would be unfair to see the Institute as an exclusively women's effort, though, for Krishnamurthi played a crucial–perhaps the central–role in giving it thrust and direction. Shanta has worked closely with Krishnamurthi and acknowledges him as her inspiration and mentor. Right from the start, Krishnamurthi gave Shanta positions of high responsibility, involving her in all aspects–acquiring land and constructing buildings, organizational structure, scientific and research achievements, fund raising, patient care and so on. Her professional life overlaps substantially with the Institute's trajectory. Indeed, it is evident from the photographs and reportage of the early decades that he was content, perhaps even preferred, to let her take the front stage while he kept a low public profile. Shanta has achieved international recognition and has won many prestigious awards.[7] I interviewed Krishnamurthi, when he was hospitalized, a month before he died in 2010 (Ganesh 2010). He expressed satisfaction that it was she and not he who won the Magsaysay award. In his functional hospital room was a picture of Ramakrishna Paramahamsa, a small statue of Jesus and a large framed photograph of a young Shanta. In the sweltering May heat, the air conditioner had been switched off. '*We will only switch it on between noon and four o'clock*', he said.

It is fair to infer that in the male-dominated field of oncology, it would have not been possible for a woman doctor to scale the heights of professional success without the platform and protection provided by a male mentor. She was totally committed to his leadership and he had complete faith in her. That the two worked together as a team, with no major differences of approach, sharing a close personal relationship has in fact been a crucial factor in the character of the institution: '*Dr. Krishnamurthi and I... you see you cannot divide him from me and me from him... we did everything together.... I learnt everything from him.*'

Dr. V. Shanta

Dr. Shanta was born in 1927, in a family with strong educational values. Even so, a professional career in medicine for a woman was unknown in the family in her generation. Shanta passed the MBBS in 1950 from the Madras Medical College. She specialized in Gynecology and Obstetrics obtaining a diploma in 1952 and an MD in 1955. She worked in the Women and Children Hospital, Madras, for two years. In 1949, she had interned briefly with the Cancer Unit of the Government General Hospital where Krishnamurthi was in charge. '*The first time I met Dr. Krishnamurthi . . . there was no doubt that it had a significant impact on me . . . his wide knowledge, his approach towards cancer, his attitude towards corruption, towards discipline. Much of what I have learnt in the early years came from him.*' She joined the Cancer Institute as a voluntary doctor in 1954. After her MD, she was offered a much sought-after permanent post in the State Service but chose to join the Cancer Institute as a full-time Resident Medical Officer in 1955. Later she became its Assistant Director, then Associate Director, becoming Director in 1980 and eventually the Chairman [*sic*] and Honorary Executive Chairman from 1997, a position which she still holds. In 1956–57, she trained in oncology in Toronto and in 1968, studied Marrow Transplantation in UK. Since 1955, she has lived in an austere flat on the upper floor of the Cancer Institute in the old complex. Despite a stroke in 2007, from which she has recovered fairly well, she continues to be immersed in the Institute's activities. Her personal and professional selves have merged seamlessly. When I asked her about her personal life and hobbies, she smiled slightly and drew a zero firmly several times in the air.

In the 1940s, Indian women studying medicine were few, but Shanta was not the only woman in her batch. She began her career in the conventional female specialization of gynecology and obstetrics. Even after it became a specialization, women were rare in the field of oncology. But then, oncology itself had just been recognized as a special field and chemotherapy had just started in the mid-1960s, in great measure due to the work of the Institute. There were few in the field at that time, male or female. And she had not come to cancer by design. *'It was destiny, will of God, call it what you like'*, she says.

What is special about Shanta is her single-minded dedication in nurturing the institution with fidelity to the goals of the founder. But, despite her involvement in all aspects, her own personal concerns lay more in the areas of access and outreach to the poor, advocacy for prevention and early treatment and organizational innovations to achieve them. She was also deeply concerned about the quality of patient care and developed protocols that went beyond treatment to holistic care. In other words, the science of oncology for her gets encompassed in what one may call an overarching 'ethic of care'.

My conversations with Shanta took place over two years and traversed several themes. Her childhood and family, schooling and medical college, critical moments in the Cancer Institute's trajectory, her views on medical education, women's contribution to medicine, and being a woman doctor, her understanding of the ethic of care, compatibility of values of science and faith, of treatment and care and many other topics. Initially measured, to-the-point, over time our interaction became more relaxed and informal; throughout, her responses were always characterized by exceptional lucidity.

Home, School, Medical College

I knew that Shanta came from a highly talented and accomplished family on her mother's side, which was traditional in some ways and unconventional in others. Her mother's brother and her maternal grandfather's brother were both Nobel Laureates in Physics. Several members of the family have made a mark in science, engineering, literature, history, music and social science.

Shanta grew up in a large joint family with her maternal grandfather, her parents and her mother's siblings. Her grandfather,

C. S. Iyer, had been a brilliant student and entered the British Railway Service as Deputy Accountant General. He was also a keen student of literature and Carnatic music and had published critically acclaimed books on the violin. His own father, R. Chandrasekhar, had already in 1888 entered St. Joseph's College in Trichy (Tiruchirapalli), read philosophy, English literature, mathematics and physics and eventually became an eminent educationist. Thus, Shanta had three generations of high educational achievements available to her as a resource, albeit all-male.

R. Chandrasekhar was a maverick, preferring to pursue education in his own way rather than conventionally as per set syllabi; he was involved in anti-colonial protests in college, approved of his son (C. V. Raman) choosing his own life partner, was an agnostic and later turned to Theosophy. For a son of a brahmin village landowner, he led an unconventional life (Wali 1991: 40). The individualistic streak has persisted through the generations.

Iyer's wife, Seethalakshmi, bore ten children, and died of tuberculosis when she was barely forty. '*Frequent childbearing had made her weak and vulnerable and that's probably why she got TB*', says Shanta, with simmering emotion. Shanta's mother, Bala, was the second oldest, already married when Seethalakshmi died. Bala and her husband, Vishwanathan, a lawyer, stayed back with Iyer to take care of the large family of six younger siblings. The two siblings next to Bala had already left home. In the early 1930s, Iyer had built *Chandra Vilas*, a complex of three adjacent houses at the end of a by-lane in Mylapore, a suburb occupied by English-educated professionals. Iyer lived in one house, which also had his personal library of thousands of books on varied subjects, in cupboards opening both back and front, in rows on the entire ground floor. The children and grandchildren lived in the other two houses. The family's cooking was done in one kitchen, recollects Malathi Ramanathan (2007), a granddaughter.

Bala spent her life nurturing her six younger siblings, and her own six children with love and dedication.

> '*She never made a difference between her children and others' children.... Friday morning 4 am she would herself give oil baths to all the children. She*

was tremendous... As we grew up we realized that she was doing so much–no, not sacrifice exactly–but so many things for others. She did not think she was doing anything special. And she got back love, affection, everything from them'.

Iyer also supported distant relatives in their times of need. His daughter Vidya Shankar, a distinguished musicologist, recalled that his two widowed aunts lived with them. Needy or orphaned nieces and nephews would stay with them on and off; he educated many of them.[8] This family ethos of matter-of-fact care giving had an impact on Shanta. The family was also exposed to more public forms of social service through their link with Sister Subbulakshmi who had set up a home for young abandoned widows.[9] Subbulakshmi's sister's daughter was married to S. Chandrasekhar, C. S. Iyer's eldest son, a Nobel Laureate, and another sister's son was married to Vidya. Thus, although the Cancer Institute's service ethos was a result of Muthulakshmi's vision, Shanta too brought her own care and service sensibilities into it.[10]

Bala had been keen to study, to become a nurse *('in her time, women in my family did not think of becoming doctors')*, she was particularly skilled in taking care of sick family members; but her education was stopped when her marriage was fixed. Shanta remembers her mother saying often, 'At least you children should study'. Bala's mother, Seethalakshmi, too had aspired for education for her daughters; although not formally educated, she had translated Ibsen's *A Doll's House* into Tamil. But her husband had conventional views. His son S. Chandrasekhar remembers feeling upset at the injustice of his father's partiality for the sons' education, giving them special private tuitions at home, while the daughters went to regular schools (Wali 1991: 54). Yet, Vidya, who was Iyer's seventh child, had a different take. Iyer's first two children were girls, married off early as per convention. By the time the second batch of daughters was born, after three sons, he grew keen on education for girls as well, although not on careers for women. Even for the boys, he was against over-specialization, and would insist on music, literature, and so on, for a well-rounded personality. Shanta remembers him criticizing some of his elite friends, who thought only of marriage and diamonds for their daughters. Buy books, he would exhort.

He insisted on music lessons: one daughter earmarked for violin, the next for veena, the third for violin and so on! Shanta too had lessons and was proficient in vocal music. Iyer had clear career plans for each of his four sons: physics, chemistry, medicine and engineering. The eldest son Chandrasekhar wanted to pursue pure mathematics, but fell in line rather meekly with his father's desire that he study physics. Balakrishnan had wanted to be a writer, but was directed by his father to choose medicine.

It was an overwhelming family to grow up in, the weight of two world renowned scientists, the authority of the grandfather, his scientific pursuit of classical music and his status as one of the founders of the Music Academy, his passion that his children excel, fuelled in part by the shadow of his illustrious younger brother C. V. Raman. When Shanta completed her school final exam at age 13, she was too young to join intermediate college. She was at home for a year, and would take down in long hand as her grandfather dictated the text of his book *The Grammar of Carnatic Music*. Shanta did not readily talk about the family, choosing to cite her school as having had the greatest influence on her. It was only when I specifically asked that she opened up about her home. It was evident that her mother had had a deep influence on her. And her grandfather? '*Well, his love of literature. His wide vision. His approach that women should be something. Both from him and from Sir CV. That you have to do something other than routine, humdrum things. To do something to make a difference*'. It was an enlightened family, she said, yet they were not relaxed about own choice or intercaste marriages. Nevertheless, some of Iyer's children did choose their partners themselves. The family was not ritualistic or orthodox, yet they held on fiercely to their cultural legacies, for example, vegetarianism. While some members identified deeply with the family ethos and stature, others were more critical, chafing for instance at the demigod status given to Chandrasekhar, as Radhika Ramnath (2008: 70), one of C. S. Iyer's granddaughters, notes. Shanta seems to have absorbed its core values of care giving and the pursuit of excellence, despite distancing herself somewhat after joining the Cancer Institute. Her closeness to Dr. Krishnamurthi and her total absorption in her work led to infrequent contact with her family.

Shanta studied at the National Girls High School in Mylapore, whose headmistress Miss Veal, an Irish woman and a theosophist, made a deep impression on her: her emphasis on discipline, simplicity, courtesy, cleanliness, how to dress, how to behave, and the insistence on certain values.

> *'It was an ordinary building. The junior classes were in thatched sheds. Sitting on the ground. Morning prayer. There were 500 or 600 children. As soon as the bell rings* [It's still happening in her mind] *there is silence, beautiful silence. I have seen that kind of silence in Satya Sai Baba's prayer meetings. From Montessori to 6th form, every one becomes silent. Then the Sanskrit teacher will come. We will all sing. Then Miss Veal will give a saying for the day. Any girl who giggles or talks . . . Finished. She has to stay back for the day.'*

By age 11 or 12, Shanta had made up her mind to be a doctor. She was inspired not by her uncle Balakrishnan who was an eminent doctor, but by her mother's life of care-giving. But there was a difference. '*I wanted to be independent, I felt that my mother had to spend too much time taking care and had nothing to fall back upon.*' Given the rather authoritarian atmosphere in the family regarding children's educational and career choices, Shanta's decision so early to become a doctor, and her resolve in pursuing it, are remarkable.

Shanta was also inspired by Chandrasekhar's wife's sister Dr. Shantasundari. When Shanta was in school, Shantasundari was in medical college and associated with Lady Dufferin's medical corps for women.[11] '*I was quite taken up with her dignified uniform, her bearing and conduct. She was not married. It is from her that I imbibed a need for being independent, not wanting to be sitting at home at somebody's beck and call.*' So departing from the family convention of opting for mathematics in the fifth form, she took up chemistry so that she could pursue medicine.

On Being a Woman and a Doctor

Everyone in the family knew about her desire to become a doctor, and there was no resistance. Even while she was at school, her aunt Vidya gave her a card in jest, with *Dr. Shanta MBBS* printed on it. When she joined medical college, everyone at home was excited. She was pampered, had many privileges. They were all a bit in awe of her and left her alone. Even her disciplinarian father Vishwanathan did not attempt to come in her way. Her younger sister got married. No one suggested that Shanta should too. It was not that marriage was ruled out for a doctor.

> '*After all, Mother* [referring to Dr. Muthulakshmi] *did get married, although late in life. So no, no, it was not that I felt that because I was a doctor, I should not marry. The circumstances were not right for me; that was all. But one thing I can say now, looking back. If you want to be a dedicated doctor, marriage is a hindrance. You cannot give everything to one, if you have another. You cannot have two dedications.*'

In those days it was not so difficult to get into medical college. Few girls applied. Medical education was free for girls till 1942, and she was in that last batch. Then for a few years, girls had to pay less, and then it became full fees. There were 12 or 13 girls in her class of more than 80.

> '*Both then and now, except for a handful, women don't think of it as a long term career. We used to say, she will get married half way and give up, depriving someone else of a seat. It is better now . . . There is no doubt that now women want to be independent, they want to work. But in their marriages, I see that sometimes they are not empowered. They buckle down. A marriage is always a gamble. The two people concerned have to make it work . . . But I see the extreme sometime. She will give up her job because he asks her to resign. Shatters your very idea of womanhood.*'

Most of Shanta's female classmates did go on to work or teach, but she did not see this fact as contradicting her earlier comment.

Shanta is not very conscious of her gender as an issue. Shanta was the first woman in the family to pursue professional education. She attributes it to her individual predilections. '*I had the need for independence. Others* [women] *may not have had that. They may have been satisfied with life.*' In a way Dr. Krishnamurthi's support and encouragement shielded her from potential gender bias and hostility from the medical world. I was keen on exploring how Shanta saw herself, as a woman and a doctor. Did she see herself as possessing special

The Healing Touch

skills as a woman? Had there been occasions in her work when she was conscious of being a woman doctor? She smiled, and said no. *'There is no time. After work, you come back, watch some TV, read and then you are dead tired and you just want to sleep. Morning you wake up, get ready and go to work. Where is the time to think? Some times when I look back, I think I should have done things differently.'*

I asked her about the differences in approach between her and Krishnamurthi. Despite their synergy, did she perceive a gender divide?

> *'Very difficult to say . . . Probably because I am a woman, I may say that emotional needs of the patient and long-term impact on family life are more important . . . We may have difference of opinion in individual cases. But the approach is not, cannot, be different. It is the approach of the institution. Yes, the institute has a distinct approach. That is, equal importance to all the three modes of treatment [radiation, chemotherapy and surgery]. Apart from that, we claimed to give individualized care. We were known for this. . . .what is the general perception about the Institute today, I can't say'.*

On Treatment *versus* Care

Shanta's emphasis on care manifests itself variously. To the question of what she enjoyed most in medical college, her immediate response was, *'patient care. . . .No, not research, etc., at that time . . . Somehow, I wanted to be with patients . . . professors do not talk to their students much about patient care . . . about morality and ethics . . . about personalized care. Dr.Sanjivi*[12] *was an exception. That was an area that interested me,'*

Living in the Institute, she used to eat the simple food cooked for patients, to make sure it was cooked properly. Vidya recalled: *'Shanta's mother used to send her something nice to eat once in a while, but she did not encourage it.'*

Krishnamurthi was moved when referring to Shanta's work:

> *'She has brought in so much care. Me, I can do surgery, I can give treatment, but I cannot give patient care the way she does. There was this young patient–Dr. Pai from Ooty. He was 22 or 23 and had cancer. I remember when he was seriously ill, he would say, "Mother, Mother, hold my hand, I am afraid of darkness." She sat with him that whole night holding his hand and in the morning, he passed*

away. That's the kind of woman she is . . . People know her. Mataji, mataji, they address her. I have technical knowledge. But for service–it is Shanta. She serves patients night and day.'

Shanta feels that her ideas on care are not to be found in practice these days.

'I hesitate to talk about it because I am unhappy about it. There is no patient-doctor relationship any longer. No empathy. No compassion. They treat patients more like a consumer article. I used to think that men are like that, women will not be like that, but nowadays even women are becoming like that. At the Institute, all our girls are very good. They will not be rude, or aggressive. They will not do something because they don't get anything from it. Even our boys are OK. But the generality is different . . . They say they need time for the family. Yes, family time is mandatory. But I say, in 24 hours, you can create time, you can always find time.'

For Shanta, the physician's role goes beyond treatment.

'Care is caring for the individual–like what you will do for your mother or father or son or daughter. It is a question of communication with the patient as soon as you see her . . . The physician has to be not just a doctor and give treatment but also care and become part of the family. All aspects of a patient's life–if she has children, then how to manage that. After the treatment, rehabilitation. Unfortunately, there is a stigma attached to cancer. Care means dealing with that as well. The physician may not be able to manage all that, but he must see to it that his set up provides for all these.

Your vibration must go…you must be part of them . . . then only they will feel . . . in the end, when they are dying, people don't want anything . . . they only want someone there to take care of them . . . when they are not so ill, when they are still OK, we have to prepare them . . . tell them half-truths . . . here we need trained counsellors . . . medical personnel are not enough . . . we don't have many such trained counsellors here . . . here the regular problems are so enormous, these smaller specialities . . . we have no time to look into them . . . the message has to be given to them very gently in a way that they will understand and appreciate . . . even when there is no hope . . . you have to give them some message of hope to sustain them . . . it is a hard thing . . .'

My question was, whether at such times she saw herself as giving something else, not just a doctor's care. She said, '*I never thought of it like that . . . all that I gave was part and parcel of medical care.*'

Science, Faith and Care

I tried to explore Shanta's own understanding of science and rationalism and how she dealt with 'care' in contradistinction to 'treatment'. Did 'faith' enter the domain of care?

> '*I am not an atheist . . . but I am not devout . . . I don't go to temples. With Dr. Krishnamurthi I have visited many temples . . . for his sake . . . yes, yes, he is quite devout . . . but now I tell him, I don't want to come . . . and he also leaves me alone . . . I don't pray . . . you would have seen in my rooms upstairs . . . there are pictures of all the deities . . . I have not bought a single one . . . people give them to me . . . I have kept a few . . . I put a few flowers and pray . . . after some days I give them away . . . if there is a problem in the Institute or a major operation . . . then I pray that things should be all right . . . never for myself . . . I visited Suchindram . . . that time the land that was given to us for the Institute had been taken away . . . of course, that Suchindram Hanumar is huge, one gets awestruck . . . I said to myself, they say, there is a God . . . then why can't somebody do something . . . I didn't tell anybody anything . . . but I prayed . . . and then we got back the land . . . there are so many small things . . . no they are not small things . . . but I find it difficult . . . to believe in karma.*'

Shanta has immense commitment to the scientific method. She implicitly conflates science with progress and scientific advancement with national pride and nation building. Nehru was personally involved in the Institute on several occasions–laying its foundation stone in 1952, opening the McConnell Radiation block nine years later, sending a cheque for the paediatric wing in 1958, arranging to waive customs duty on the import of the Cobalt 60 unit in 1961. That both Shanta and Krishnamurthi were united in their admiration for him is symbolic of the Institute's embeddedness in the spirit of the Nehruvian era.

Shanta is not opposed to alternate health and medical systems like Ayurveda or Siddha, but insists on rigorous scientific testing before they can be accepted. She refrains from commenting on their efficacy, since she has not made any systematic study, she says. For her own patients who often ask whether they can try out this or that remedy, she leaves it to them, says she cannot take responsibility, and urges them not to discontinue the treatment at the Institute.

She is sceptical about cases of 'spontaneous remission'–where the cancerous cells disappear for no known reason. She attributes them to the genetic propensities of individuals, even though she says the findings of genetic oncology are still not conclusive. But she agrees that hope and faith, and a positive attitude do seem to have some benefits for patients. Role of will to live, hope and religious beliefs?

> *'I don't know, but I can say that someone who is more positive and goes with the treatment does better . . . whether it helps in cure or long term benefits . . . hard to say . . . I don't think I should comment on it because I have not studied it in depth . . . majority of people who come do have a lot of faith . . . so what sees them through? . . . one of them is faith, number two is family support . . . three is medical support . . . compassion that they receive from medical professionals . . . confidence in the doctor–significantly on the attitude of the doctor.'*

Gender and the Ethic of Care

Feminist debates on care ethics furnish us with reference points to understand Shanta's engagements with cancer treatment *versus* cancer care. Care ethics as a distinct moral theory emerged in the 1980s, and the works of psychologist Carol Gilligan (1982) and philosopher Nel Noddings (1982) were crucial in this. Both found that traditional moral theories had male bias, and proposed the 'voice of care' as a valid and compatible alternative to the 'justice perspective' of liberal human rights theory.

The empirical fact is that care work all over the world is done mainly by women–whether care giving in the private domain of home, voluntary work in the community or paid work in the market. For that very reason, as a profession, it is poorly paid and has low status. Unpaid care work at home and for the community, while romanticized and sentimentalized, does not empower women materially. The 'double burden' of housework and work outside the home, and the difficulties of getting men to share housework are well-established research findings. Yet care work is of crucial importance for the well-being of society at large and needs to be given cognitive status.

If the complex feminist debates on the subject are condensed, then it can be seen that, these have attempted to valorize care ethics as contributing to political theory and global relations and applicable to a number of moral issues and ethical fields including valuing women's care work, caring for animals and the environment, bioethics, and public policy. At the same time, feminists question the notion of care as an essentially female virtue and critically interrogate the empirical and symbolic association between women and care, focusing on its power-related implications. They put individual actions into social contexts, and emphasize the differences among women in orientation towards care (for a concise account of recent debates see IEP 2011).

Gilligan (1982) critiqued the classic account of Lawrence Kohlberg that defines moral development as progressively universalized and principled thinking. She proposed that an alternative voice to the masculinist preoccupation with autonomy and justice could come from relational thinking with a care ethic at its core. This 'different voice' is associated largely with women, but not exclusively so. Noddings (1982), Virginia Held (2005), Kittay and Feder (2002), Sara Ruddick (1995) and others have fleshed out and nuanced this position, discussing concepts like 'feminine ethics', 'maternal thinking', 'feminist morality' and so on. Joan Tronto (1993) has taken this debate further by exploring the intersections of care ethics, feminist theory, and political science. She identifies the traditional boundary between ethics and politics as one of three boundaries, which serves to stymie the political efficacy of a woman's care ethic (the other two being the boundary between the particular and the abstract/impersonal moral observer, and the boundary between public and private life). Together, these boundaries obscure how care as a political concept illuminates the interdependency of human beings, and how care could stimulate democratic and pluralistic politics. Care work is done predominantly by women but also those who are economically and socially marginalized. Advantaged sections who can purchase this care, manage to get into positions of 'privileged irresponsibility' by avoiding having to do hands-on care taking. In cultures of the advanced industrialized West, the valorization of individual autonomy is achieved at the cost of a denial: of women's unpaid care work on which society and all its public institutions depend (Fineman 2000: 14). Women's dependence on breadwinning husbands is labelled as 'dependence' but men's dependence on care-giving wives is conceptually ignored (Ganesh 2005b: 147).

Shanta made her way in the male world of oncology becoming, in some ways an' honorary male'. It is intriguing that at the Institute she is referred to as 'Chairman', whereas Muthulakshmi Reddi was designated as 'Chairwoman' of WIA. Through her competence and efficiency, Shanta ensured that her being a woman could not be cited negatively. In the larger canvas of the Institute's struggles, gender is not her priority. Yet her own actions in the sphere of care giving do draw from traditional resources—specifically her mother's role as care giver, even though she has transmuted it from the narrow kinship arena onto a large setting with a universalist thrust. In this sense, we could see Shanta as bridging the two stages of caring, 'caring-for' and 'caring-about' identified by Noddings (1982), that is, actual hands-on application of caring services versus a state of being whereby one nurtures caring ideas or intentions.

The feminine care ethic is based on care as a woman's duty with an emotional overlay, obscuring the labour that goes into it (which for instance facilitated the men in Shanta's family to become world famous scientists). As an obverse, the onus of action–to rebel against such ascriptive roles, is also placed by Shanta on women, rather than on structural change. When women doctors do not take their work as seriously as they should, or when women complain about domestic violence without walking out, she sees it as their failure to resist. But one should not forget that Shanta's grandmother's translated *A Doll's House*. It must have had some reverberations among the women in the family. Superimposed on all this was the 'scientificity' Shanta acquired in childhood from her scientist relatives and then from her own medical training. Her thinking, speaking and doing follow different modes. It is in her 'doings' that we see her care ethic in full bloom. Shanta's is a rather complex and shifting position interleaving male, feminine and feminist sensibilities. This is indeed a shadowy area of ambiguity for a person with such clarity in articulation. Muthulakshmi's own outreach activity too was not just a 'womanist' concern but a result of her being a particular kind of woman embedded in a certain era and ethos.

A macro institution concerned with health and medicine can be analysed at many levels–in terms of deliverables, efficiency, coverage, outreach, affordability, economy of scale, technical capabilities, contribution to national development, adaptation to changing external environment and so on. But when one looks for the well spring of motivation, the source invariably lies in the domain of individual lives and personal relationships. The sorrow of bereavement of a sister, stormy love of a son and the fierce loyalty of a comrade ignited the passionate commitment of the three protagonists of this story. Their combined energies went into the making of a unique institution. As feminists, we are ever alert to the power of the intimate domain. Behind everything lies a personal story.

NOTES

1 There were, of course, other notable doctors at the Institute, as also a team of dedicated nursing, administrative and other staff and volunteers, who played important roles.
2 In 2010, the Avabai Wadia Archives for Women (AWA), at the Research Centre of Women's Studies, SNDT Women's University, gave me a fellowship to put together an archive of the Cancer Institute, which provides the base for this essay. The specific methods of archiving included: interaction and interviews with Dr. Shanta and observations at the Institute over two years (2010–11), structured interviews and open-ended conversations with doctors (from the Institute and outside it), patients, members of the governing board, and volunteers; relatives and friends of Dr. Shanta. The records of the Institute dating from the 1930s and other secondary literature were important sources. These records, photographs, voice recordings of the interviews done by me are now part of AWA, accessible to researchers and students.
3 The Golden Jubilee volume of the Institute (*Five Decades*) has rich information on the details. The Dr. Avabai Wadia Archives has copies of the Annual Reports and other original documents on the subject.
4 Direct excerpts from my interviews are given in italics.

5   She resigned in 1930 as soon as she heard of Gandhi's arrest during the Salt Satyagraha.
6   The organizations included the Madras Presidency Muslim Ladies' Association, The Women's Indian Association, Sri Sarada Ladies Union, The Young Women's Christian Association and Madras Branch of the All India Women's Conference.
7   To cite just a few, Padma Shri (1986); Padma Bhushan (2006); Nazli-Gad-El-Mawla Award of the International Network for Cancer Treatment and Research for Outstanding Contribution to Cancer Control in a Country with Limited Resources (2002); Ramon Magsaysay Award for Public Service (2005); and several lifetime achievement awards and honorary doctorates.
8   To aid the reader in following the threads of Shanta's family, I have listed below the family members referred to in this article in the order in which they appear, along with their relationship to Shanta.
    C. S. Iyer: maternal grandfather
    R. Chandrasekhar: maternal great grandfather, C.S. Iyer's father
    C. Raman: maternal granduncle, C. S. Iyer's younger brother
    Seethalakshmi: maternal grandmother
    Bala: mother
    Vishwanathan: father
    Malathi Ramanathan: cousin, Balakrishnan's daughter
    Vidya Shankar: maternal aunt
    S. Chandrasekhar: maternal uncle
    S. Balakrishnan: maternal uncle
    Radhika Ramnath: cousin, daughter of C.S. Iyer's youngest son
9   R.S. Subbulakshmi or Sister Subbulakshmi as she was known, was a child widow from an orthodox brahmin family who educated herself and worked for the welfare of women, setting up the Sarada Ladies Union in 1912 and Sarada Illam, a home for widows. She was awarded the Kaiser-e-Hind by the British government.
10 Her own comment on my interpretation of the source of her service ethos was that it was really derived from her medical training and the Hippocratic Oath.
11 Lady Dufferin the wife of the British Viceroy, led a campaign to improve medical care for women in India. She started the National

Association for supplying Female Medical Aid to the Women of India ('The Countess of Dufferin Fund') in 1885. The association trained women doctors, midwives, and nurses, to improve the treatment of Indian women in illness and childbirth.

12 Dr. K.S. Sanjivi was a renowned physician whose concern for affordable health care led him to set up the Voluntary Health Service that became a model in its field.

REFERENCES

Anandhi. S. 1991. 'Representing Devdasis: *Dasigal Mosavalai* as a Radical Text', *Economic and Political Weekly* 26, 11 and 12, Annual Number: 739–46.

Basu, Aparna. 1987. *The Pathfinder: Dr. Muthulakshmi Reddi*. New Delhi: A.I.W.C.

Basu, Aparna, and Bharati Ray. 2003. *Women's Struggle: A History of the All India Women's Conference 1927–2002*. New Delhi: Manohar.

Bennett, M. Judith. 1989. 'Feminism and History', *Gender and History* 2, 3: 251–72.

Cancer Institute. 2004. *Five decades of the Cancer Institute (WIA)*. Published on the Occasion of its Golden Jubilee (February 2004), Chennai.

Ganesh, Kamala. 2010. 'A Doctor and an Institution-Builder', *The Hindu* (July 7). http://www.thehindu.com/opinion/op-ed/article503098.ece (accessed 18 July 2012).

———. 2005a. *Illness as Metaphor*. Tehelka, 10.9.05: 22–23. http://www.tehelka.com/story_main14.asp?filename=Cr091005Illness_as.asp (accessed 18 July 2012).

Ganesh, Kamala. 2005b.'Made to Measure: Dutch Elder Care at the Intersections of Policy and Culture'. In *Care, Culture and Citizenship: Revisiting the Politics of the Dutch Welfare State*, edited by Carla Risseeuw, Rajni Palriwala and Kamala Ganesh, Amsterdam: Het Spinhuis: 116–58.

Gilligan, Carol. 1982. *In a Different Voice: Psychological Theory and Women's Development*. Cambridge: Harvard University Press.

Fineman, M. 2000. 'Cracking the Foundational Myths: Independence, Autonomy and Self-Sufficiency', *American University Journal of Gender, Social Policy and the Law* 13: 13–30.

Harding, Sandra, ed., 1987. 'Introduction: Is There a Feminist Method? In *Feminism and Methodology*. Bloomington: Indiana University Press: 1–13.

Held, Virginia. 2005. *The Ethics of Care: Personal, Political and Global*. New York: Oxford University Press.

IEP. 2011. 'Care Ethics'. *Internet Encyclopaedia of Philosophy*. http://www.iep.utm.edu/care-eth/ accessed 18 July 2012.

Kittay, Eva Feder, and Ellen K. Fede, eds, 2002. *The Subject of Care: Feminist Perspectives on Dependency*. Lanham, VA: Rowman and Littlefield Publisher.

Matthews, Jill. 1986. 'Feminist History', *Labour History* 50: 147–53.

Noddings, Nel. 1982. *Caring: A Feminine Approach to Ethics and Moral Education*. Berkeley: University of California Press.

Ramanathan, Malathi. 2007. 'The Cancer Institute (W.I.A.), Chennai: Its Unrelenting Fight against Cancer' International Conference on 'Quest for Excellence: Great Universities and Their Cities–Calcutta, Bombay and Madras', Organized by the Department of History, University of Mumbai, Mumbai, January 2007.

Ramnath, Radhika. 2008. *Family Portraits*. New Delhi: Privately Published.

Reddi, Muthulakshmi. 1964. *Autobiography*. Madras: M.L.J. Press.

Ruddick, Sara. 1995. *Maternal Thinking: Toward a Politics of Peace*. Boston: Beacon Press.

Stri Dharma, 1928. vol. 11, Madras: Women's Indian Association.

Tronto, Joan. 1993. *Moral Boundaries: A Political Argument for an Ethic of Care*. London: Routledge.

Wali, Kameshwar. 1991. *Chandra: A Biography of S. Chandrasekhar*. New Delhi: Viking-Penguin India.

19

Body, Reproduction and Technology: Local Subversions and Global Regressions

Chayanika Shah

> *Robert Edwards is awarded the 2010 Nobel Prize for the development of human in vitro fertilization (IVF) therapy. His achievements have made it possible to treat infertility, a medical condition afflicting a large proportion of humanity including more than 10 per cent of all couples worldwide. . . . Infertility—a medical and psychological problem . . . For many of them (couples), this is a great disappointment and for some causes lifelong psychological trauma. Medicine has had limited opportunities to help these individuals in the past. Today, the situation is entirely different. In vitro fertilization (IVF) is an established therapy when sperm and egg cannot meet inside the body.* (Nobel Prize 2010)[1]

In the last 110 years of the declaration of the Nobel for Physiology or Medicine no other intervention in reproduction has been deemed worthy of the prize. The knowledge of the hormonal cycle that controls ovulation, and the ability to control it with artificially produced chemicals—the key to the discovery of IVF as stated in the advance information for the 2010 prize (quoted above)—did not deserve any such prize.

In this chapter, I wish to address the mindset of science and technology through the tale of this reproductive cycle and the technological interventions in it for contraception or infertility assistance. An overview of this process shows how infertility, overfertility or unwanted fertility get termed medical conditions and quick fix medical, interventionist solutions for these conditions are lauded and simultaneously the social and psychological implications and solutions are underplayed and negated. I use this example to illustrate the complex terrain of interwoven threads of gender, body, society, science and technology and in the process help critique the ways in which established and practised science looks at women and their bodies. This critique is essential for our engagements with both, feminist theory and action. It is also crucial for the growing Feminist Science Studies (FSS) scholarship because the critique helps identify paths for the future. There is no answer to what kind of science will emerge but there definitely is a roadmap for the ways in which this pursuit needs to be made.

Talking of IVF and Infertility

First of all, what is it that the Nobel Committee awarded? We need to understand what is meant by both infertility and IVF therapy for treatment of infertility. The *Encyclopaedia Britannica* (online, 2012) says that, infertility is the 'inability of a couple to conceive and reproduce. It is defined as failure to conceive after one year of regular intercourse without contraception or the inability of a woman to carry a pregnancy to a live birth.' This failure can have many reasons. The cause of the infertility could be physiological: blocked fallopian tubes in the woman due to malnutrition, infection and a myriad other such reasons leading to the egg being released but unable to meet the sperm inside the woman's body; lack of ovulation, due to causes like repeated use of hormonal contraception, emotional or other stress, polycystic ovarian disease (PCOD), diet imbalances or some other such reason leading to the absence of the egg itself; azoospermia or lowered motility of the sperm as a result of occupational or environmental reasons leading to the sperm's inability to meet the egg in vivo (inside the woman's body). Or it could be the fairly frequently occurring, totally unexplained, infertility which means that the egg and tubes are all right, the sperm is very much there, individually each of the partners is potentially fertile and intercourse is taking place but for some reason there is no pregnancy.

So treating infertility should actually mean opening up the blocks, creating bypasses for the tubes, making sure the eggs mature, improving the quality of sperm, getting rid of the infections, reducing the intake of hormonal contraception, basically figuring out the reason for the lack of conception and addressing it. And there are medical solutions for infertility that do all this. Indeed, this is what we expect medical solutions to do. The IVF therapy, however, is a medical solution that does nothing for these physiological causes of infertility. It does not address the cause at all. At the end of the treatment each individual partner, or the couple, remains as infertile or fertile as before. It just makes sure that the child is born with the germ cells from the couple. So the therapy actually helps in creating a baby with the required genetic material by making the egg and the sperm of the couple meet outside of the faulty body, in a petri dish, and starting the process of creating the baby there instead of in the fallopian tube.

If the man's sperm is not good enough, the medical team will provide assistance by selecting the few sperms that are capable of fertilization from the millions that are ejaculated. The sperm might not manage to fertilize the egg on their own but the external medical procedure can definitely help. If the egg is not getting formed in the woman's body the therapy will make do with an egg from an oocyte donor and use the man's sperm and make the couple fertile by either inserting the embryo back into the woman who is part of the couple or any other selected by the couple (a surrogate).

So it does not necessarily help conceive, it helps create a baby of the requisite make. Thus, over the years, we have developed a medical method useful for some couples who want to have genetic babies of their own—a medical solution to a couple's problem. Interestingly, the therapy helps address male infertility also through extensive medical intervention in women's bodies. 'The development of intra-cytoplasmic sperm injection (ICSI), in which a single sperm is microinjected into the cytoplasm of the mature egg, represented a technological breakthrough, making it possible also to treat many categories of male infertility' (ibid.). This is probably a first in medical procedures where the intervention for one person's problem is completely in someone else's body!

It is the idea that fertilization of the human egg could happen in vitro (outside the body) that got Edwards the Nobel Prize. The Nobel committee has appreciated the achievement of taking control of the process of conception, making it happen 'artificially' outside the 'natural' environment of the human body. This is a feat that seems to fascinate modern science and technology practitioners and is applauded as progress and development every time.

Technology of Choice or Control

In a way infertility is a medical condition because it involves biological processes but it is understood more as a psychological condition as the Nobel citation itself states. And like modern medicine treats other psychological problems with only chemicals; it does the same here. Not having a child is psychologically traumatic because of social reasons. Indeed, infertility is a completely social problem. What the definition of infertility does not say is that in most societies the couple is presumed to be married. Unmarried heterosexual couples in active sexual relationships would wish infertility on themselves as for them fertility could lead to a psychological and social condition. The solution hence does not have to be medical. Adoption or living happy child-free lives could be perfect, normal solutions (only this will not bring forth a Nobel prize). Modern medicine offers a way to deal with this social problem without touching the socio-structural core of the problem.

Talking of prenatal diagnostic techniques, Anita Ghai and Rachna Johri (Chapter 5, Volume 1, 2015), state that: 'Although technological developments enhance the "sense" of choice, in reality they tend to push decisions in a predictably socially constructed direction. In contemporary India, technology of prenatal determination of foetal characteristics has disadvantaged both girls and the disabled.' My premise is that reproductive technologies do the same, give a sense of choice to the individual concerned while sealing the fate of many others by reconsolidating the normative. IVF offers married couples the choice of having a child with their own genetic material but at the same time it underlines the normalcy of living a life in a blood-and genes-or-marriage-linked family.

IVF has potential for subversion and in theory can help any person have a child, even those socially and biologically not allowed to have children like single persons or same sex couples. But use and access to technology is always through the social fabric. Technology makes it seem like a choice is available but all technology is controlled by dominant power structures and so rarely does society allow a subversive use of technology. Hence, technology cannot ever provide a fair solution to a social problem. And this holds true also for the social problems of inequality, discrimination and injustice. Technology may sometimes seem like a leveller but it can never be a solution. History provides many examples. By their very existence, machines should have made muscle strength irrelevant in industry but at no time did well-paid jobs in the mechanized or even computerised industry ever come to women supposed to be physically weaker. Every turn in the growth of industry has meant removal of women from factories or their being relegated to lesser paying, more routine, piece meal, home based, unorganized jobs.

The story of hormonal contraceptives is similar. The discovery of artificial hormones and oral contraceptives more than half a century ago, allowed women to have the freedom of penetrative heterosexual intercourse without the fear of pregnancy. Oral contraceptives prevented pregnancies but freedom of any kind was a far cry. Much has been written on the contribution of the pill to the 'sexual revolution' of the 1960s in the USA in particular. The pill and an easy access to contraception for all women thereby was assumed to be the solution for the big problem of unwanted conception and lack of access to safe and legal abortions. The sexual revolution was not triggered by the pill but some single women and many married women did benefit.[2]

Some women's stories tell us, however, that from their point of view it was not the sort of freedom that they wanted. Even in situations where women chose to use the pill, there were many unresolved issues. Access to the pill meant no access to the possibility of saying no to sex. So while it did allow them the freedom to choose to have penetrative sex with partners of their choice, whenever they wanted it, that their relationships did not get liberated in the same ways meant that their male partners did not heed their refusal any more. Excuses of possible pregnancy and of being found out were no longer available. At the same time it reduced possibilities of conversation and negotiation around use of barriers and other methods of birth control like coitus interruptus. Getting pregnant was seen as a woman's problem and a pill a day seemed like an easy, modern and also efficient solution. Scientific and effective medicine apparently took on the task of people's sexual lives distancing them more from their bodies and their relationships.

The introduction of domestic technology had successfully improved the conditions of housework for women without liberating them from being solely responsible for housework. Similarly, the pill made life a little simpler for women who were subjected to repeated pregnancy and childbirth as they did not have access to methods of birth control and safe abortions. It did not take the onus of using contraception away from them. It did not change societal controls over their sexuality and reproduction. It did not question the dominance of the peno-vaginal penetrative heterosexual act of sexual pleasure over all others and thus did not help question or alter the equation made between sexual pleasure and reproduction.

There was severe opposition to the pill from conservatives because it allowed sex without the possibility of conception and birth thereafter. It went against the tenets of religion and messed with the existing notions of social morality. And yet research in hormonal contraceptives continued due to the hard campaigning by women's rights activists like Margaret Sanger. In 1950, when Sanger was in her eighties, she underwrote the research necessary to create the first human birth control pill. She raised $150,000 for the project, and in 1960 the first oral contraceptive, Enovid, was marketed in the United States (BWHBC 2012). Sanger could not bear to see the plight of women being forced to have children, unable to access safe abortion because of the rigid diktats of church and society. She had been a tireless crusader for birth control since the early part of the century. But there were other agendas at work, which made the pill the most researched contraceptive method and also popularized it all over the globe.

'Something Like a War'

There was tacit support to the pill and research in contraception in the USA and other 'developed' nations because it was seen as a means of controlling population. As one researcher David Allyn (2001) says 'In his State of the Union address of 4 January 1965, President Lyndon B. Johnson cautiously announced his plans to promote the use of birth control abroad. "I will seek new ways", he told Congress, "to use our knowledge to help deal with the explosion in world population and the growing scarcity in world resources". This statement of support for birth control was not a bold one, but it was the first ever by a sitting president.' This reflects the growing global politics of 'population control' which has become prominent and holds sway over popular understanding across the globe today–A politics that is based on the belief that to deal with limited resources on the planet the only solution possible is of controlling the number of people on the planet.³

Despite continuous evidence of the lopsidedness in consumption patterns and the ineffectiveness of force and coercion in achieving families, communities and nations of the 'right size', the belief in this logic is very deep seated. It also stems from the conviction that the poor, the marginalized and women do not know what is best for them and that progress and development as defined by the long history of modernism is the only definition possible. Population control is hence inherently eugenic in character. The effort is not only of keeping the numbers right but also of balancing the numbers of the 'right' kinds of people, racial characteristics, genders, abilities, and so on.

This mindset is aided by the current model of development which believes in uncontrolled consumption of global natural resources by the few who can afford it at the cost of drastically reducing the numbers of many who really cannot afford the same kind of consumption (see Bandarage 1997; Hartmann 1995; Rao and Sexton 2010). These population control policies are by their very nature coercive and eugenic and thus targeted at certain populations across the globe. Family planning programmes implemented especially in the South were the local face of global politics. They acted locally through providing contraceptive methods that control and regulate the number of children born, as per global targets set rather than the need of the individual. And research into the medical control of reproduction was an important tool in this effort.

Abetting this has been the pharmaceutical industry which has seen a large market in this need for freedom from unwanted pregnancies. For an industry that caters usually to the ill and ailing, this market of healthy persons requiring the drug for such long spans of their lives obviously meant good business. All medical research is anyway funded by the pharmaceutical companies whose interest, of course, is creating a market and keeping it alive. So contraception is marketed continuously and changed all the time so that people feel new choices are emerging even though these may be just repackaging of the old. There is ample evidence for the ways in which vested interests affect research through fudging data and trials, flouting research trial norms, biased interpretation of data while claiming neutrality and objectivity, using people's vulnerabilities to make more profits–the list can go on (Forum for Women's Health 1993).

And so research in hormonal contraceptives continued from the time that the knowledge of the hormonal cycle was acquired. This knowledge in itself does not necessarily mean that the next step has to be controlling it. With the mindset described above, however, manufacture and consumption of artificial hormones to manipulate and 'regulate' women's periodic fertility cycle became the path of progress and growth for hormonal contraceptives. Hence, regulation is done through disruption of the cycle so that no eggs are released, thus, suspending the woman's fertility in some temporary or permanent ways. This research was dictated by the needs of international global politics and the demands of the population control and pharmaceutical industry, combined and matched completely with the notion of progress of a particular model of science. Their interests were fairly clear–controlling population through reducing births, maximizing profits and obtaining a degree of control over women's bodies.

The pill is claimed to be the most researched contraceptive available in the market but the chemical composition of pills have been experimented by a trial and error method all along. It was introduced in a rush and then followed up with large-scale clinical trials when many women reported all kinds of adverse effects after using it. As the Boston Women's Health Collective (2011 website) documented in *Our Bodies, Ourselves* said, 'The early pill formulations raised concerns about blood clots, heart attack, and stroke, spurring exhaustive research on oral contraceptives beginning in the 1960s and 1970s. Since then, new formulations of the Pill with low-dose hormones have been introduced, and today's pills contain about one-eighth to one-tenth of the estrogen in early pills.' Concentrations and different compositions of the synthetic hormones were tried. As more women complained of adverse effects, newer combinations became popular. So the market continued to woo the individual consumer of the pill by bringing in newer and minutely different combinations every few days.

'Just a Pill a Day' to 'No Worry for Five Years'

The other direction that research took was in finding ways of making the contraceptive effect longer lasting. A pill a day was considered a nuisance worth doing away with. Doctors and scientists apparently worked towards making life for women simpler by releasing them from the daily trouble of remembering to take the pill. The focus of this research has hence been on making the methods such that the delivery of the hormones into the women's cycles did not require women to take the pill. The concern was not so much women's experience of the drugs and their side effects, but the need to find methods that could be controlled by the provider. The aim was not to find contraceptive methods that would suit women; it was to look for methods that would ensure that women would use the method irrespective of whether they liked it or not.

The sole criterion for success of a contraceptive was its efficacy. Women's experiences of the drug did not count. Their agency in making a choice of what method to use, when and for how long mattered even less. Indeed, if we look back at the development of hormonal contraceptives in the past six decades, we find that the argument of overpopulation has foreshadowed the development and distribution of contraception all these years. The path that this research has taken, therefore, moves towards developing more and more invasive, long acting, provider-controlled methods. It is not a coincidence that this is also in keeping with the overall attitude of control advocated by science and advocated and pushed by researchers and doctors as well.

The first such method to be prepared and introduced in this phase of the development of the technology were injectable contraceptives whose effects last for a period of two to three months. A high dose of the drug is injected at one shot and its concentration and effect is expected to wear off after the stipulated time period. The injection itself is obviously irreversible. No one can do anything if the injection results in severe menstrual irregularities (bleeding or absence of it or both), abdominal pain or discomfort, weight changes, headache, weakness or fatigue, hair loss or nervousness. And these adverse effects of the injections have been reported throughout the history of the use of the drug from the time of its clinical trials. Animal studies had shown, right at the beginning, that there was an increased risk of cancer when exposed to this drug. Later long term studies with human subjects have shown that there is serious risk of reduction in bone density for users of the injectable Depo Provera.

The injectables were first tried in the late 1960s, the peak period of women's health movements in the USA. The results of animal trials and women's experiences with the injectables during clinical trials were well documented by women's activists and groups. On the basis of the serious adverse effects recorded, women's health activists protested against such contraception very successfully.

In India, the injectable Net En was first tried way back in the 1980s. Women's groups in India drew upon the material generated by the women's groups in the North, and protested against the trials being conducted in India. Through the joint interventions of groups across the country, a case was filed against the introduction of these injectables into the national Family Planning programme and the trials stayed (Sathymala 2001). The efforts were successful in stalling the introduction of the contraceptives into the family planning programmes. The case was in the Courts for 15 years. As Saheli a women's group in Delhi, who were co-petitioners in the case say: 'At the closure of the case, the Court directed that Net-En could only be introduced where adequate facilities for follow up and counselling are available.'

Upjohn, the company that first tested Depo Provera, applied for FDA approval in the USA in 1973 but this was rejected due to various problems with the injectable in the animal trials and lapses in the way that the clinical trials were conducted. The company kept applying and in the meanwhile also managed to subject more than nine million women across the globe to the drug. Finally the FDA granted approval in 1992 giving the absurd reason that animal studies were not really indicative of what would happen in human beings. Whatever the pressure the FDA acted under, this led to a greater approval to Depo in different countries. In India, the Drug Controller of India approved its use in 1994 for 'social marketing'. The injectables could not be introduced into the family planning programme but they are being made available as over the counter drugs and free of cost through large NGOs that help run some of the family planning clinics in lieu of the government.

In North America the battle continued. As more and more women used the injections or were forced to use them, they realized that the threat of reduction of bone density was not a minor adverse effect. They saw that the information that the company (now Pfizer) shared was inadequate and so a class action suit has been initiated in Canada in September 2005; the Supreme Court of Quebec authorized it in May 2008 and in July 2010 a class action has been instituted against Pfizer Canada Inc. and Pfizer Inc. (Lepointe 2010).

Meanwhile the worldwide protest movements against the high dose injectables changed the research agenda to create a method that released very small concentrations of the chemical into the bloodstream but managed to retain the provider-controlled character and long-term action. These were slow release implants which were long lasting (five years) and were propagated as being less problematic than the injectables. 'The dose is even less than a pill' the doctors said and hoped that the problems would go away. But the problems that women complained of did not decrease. There were additional issues because the implants had to be removed after the period of action. In the phase 3 trials in India, a large number of women were lost to follow up, which means they never came back to get the implants removed and carried the possibly carcinogenic silica implants in their bodies for unknown periods of time (Chayanika, Swatija, Kamaxi 1999).

Removal was a surgical procedure which was claimed to be simple and safe but did not actually turn out to be so. If women wanted to remove the implants because it did not suit them, they could not do so as they were dependent on the provider to do this and for various reasons the providers refused to remove the implants. In India due to the pressure of completing the trial with a given number of respondents, the investigators refused to remove them. We have recorded women speaking of their trauma of having to continue with the implants because the providers refused to remove them. One woman from Vadodara who was part of the clinical trial told us that she went to the doctor some twenty times to get it removed and finally managed to get it out after a lot of crying, pleading and threatening.[4]

In Indonesia the story was slightly different because the implants were introduced in a rush. Here the providers had got the training to insert the sub-dermal implant but had not been trained in removal as it was assumed that this would happen only after the full period of five years. As Vimal Balasubrahamanyan (1993: 088) cites from a study assessing 'acceptability' of Norplant in the Dominican Republic, Egypt, Indonesia and Thailand,

> The gist of it is that in all four countries there were reports that removal on demand did not occur to the 'satisfaction of the user'. In Thailand doctors insist that users should continue the method because it is costly. Thai women getting Norplant are routinely told that because it is a long-term spacing method, it will not be removed for 'minor side effects'.

In the USA black women and other people of colour were the recipients of implants as they were the 'South' in the North. INCITE! Women of Colour Against Violence gives information on the use of Norplant and Depo Provera in the USA under the head of 'dangerous contraceptives':

Now that coercive sterilization is less acceptable, long-acting hormonal contraceptives like Norplant and Depo-Provera have become the primary tools against 'overpopulation'. Several women's organizations, such as the Black Women's Health Project, the Native American Women's Health Education Resource Center, the National Latina Health Organization, and the National Women's Health Network, have opposed these methods as appropriate forms of contraception. Several state legislatures have considered bills which would give bonuses to women or public assistance for using Norplant. The *Philadelphia Inquirer* ran an editorial suggesting that Norplant might be a useful tool for 'reducing the underclass'. Judges haven even required women convicted of child abuse or of drug use during a pregnancy to use Norplant.

Women's experiences of the implant are far from satisfactory and none of the patient information provided prepares women for the kind of problems that they face. As a result a class action suit was filed in November 1993 in Louisiana against Wyeth Laboratories, the makers of Norplant and in 2011 the petitioners arrived at an out of court settlement of $29.5 million with the company. As part of this, all women who got implants inserted and removed in Louisiana and suffered from listed side effects could claim for compensation (Silverman 2011).[5] In India the state has tried time and again to introduce the five-year hormonal implants but has not succeeded so far, partially due to pressure from the women's groups but also perhaps because five years is too long a period for its use as a spacing method; or because they were waiting for the class action suit against Norplant in the USA to be settled. In 2004 Indian Council of Medical Research started a 17-centre, Phase III study evaluating Implanon, a single-rod contraceptive implant made by the Dutch manufacturer Organon, which provides contraceptive protection for three years. The results of this study are still awaited (Sarojini 2010).

This is the chequered journey of hormonal contraception–an intervention whose beginnings lay in the simple understanding of the hormone cycle and the possibility of intervention through artificial hormones. The attitude is of manipulating the cycle while looking at women's bodies in reductionist ways, mechanizing the whole process of reproduction and allowing control to shift completely into the hands of the provider of the technology or the biochemical system that takes over the usual functioning of the body. In development and popularization of all of these methods, women's bodies and their abilities to reproduce have been manipulated and controlled at the cost of their health and well-being.

Research in Contraception: Acquiring Knowledge or Control

An argument may be made that this is all about the political use of technology and has nothing to do with the nature of medical science or technology itself. This, however, is a simplistic and blinkered way of looking at the agendas of knowledge production. The question is not about use and misuse of technology, this seems to be the way in which reproduction, bodies and interventions have been understood by modern medicine. As Vandana Shiva (1988: 23) says while critiquing the view that science is just a discovery of facts about nature,

> .... the context is determined by the priorities and values guiding the perception of nature. Selection of the context is a value determined process and the selection, in turn, determines what properties are seen in nature. There is nothing like a neutral fact about nature, independent of the values shaped by human cognitive and economic activity. Properties perceived in nature depend on how you look at them, and how you look depends on the economic interest you have in the resources of nature. Looking does not create properties, but it definitely creates conditions for their perception.

Carolyn Merchant (1990) provides one of the earliest critiques on this aspect of modern science. In her critique of the historical growth of modern science. Evelyn Fox Keller (1985: 54) builds from this and other works to say, 'In true Baconian idiom, Joseph Glanvill adds that the function of science is to discover "the ways of captivating nature, and make her subserve our purposes". The goal of the new science is not metaphysical intercourse but domination, not the union of mind and matter but the establishment of the "Empire of Man over Nature".'

Body, Reproduction and Technology

An overview of the general 'progress' in understanding of reproduction and developing of all contraceptive method shows how development and progress have basically meant developing methods that are invasive and controlling of the natural cycles and processes of the body. The first knowledge is of the fact that conception happens when the ejaculate from the man enters the vaginal canal of the woman; hence, the first method used is coitus interruptus where the man makes sure to not ejaculate inside the woman's vaginal canal.

The earliest known contraceptive devices similarly are the simple and obvious barrier methods which prevent the ejaculate from entering the vagina by creating a barrier of a condom worn on the man's penis or a diaphragm placed inside a woman's vagina.[6] Obviously these methods did not alter anything in the bodies and physiological processes. If at all they altered something, it was the process of sexual interaction between people. But they are looked down upon because they are considered inefficient with a failure rate of 2 per cent (Chayanika, Swatija and Kamaxi 1999) and also because they are dependent on the active participation of the people involved.

The next methods that emerged were invented in the early part of the twentieth century. These acted post-conception by placing something within the uterus and thus preventing the implantation of the embryo into the uterus. These were the intrauterine devices which came in various shapes and sizes to ensure full occupation of the uterus. This was the first device that was placed inside the body and that too for long periods of time. Many variants of the IUDs in terms of shapes, sizes and materials were made and are still available. While the device stays in the body it obviously affects other regular functions of the body like contractions of the uterine walls during menstruation. With the IUDs, contraception was out of the direct control of the woman, was long acting, was inserted in the body and intervened in the regular processes of the woman's body irrespective of whether there was possibility of penetrative sexual intercourse leading to conception.

Next in this brief hundred-year history of modern medicine and its understanding of contraception is the knowledge of the hormone cycle and the possibility of disrupting it through chemicals resembling hormones. The perfectly well-balanced cycle was disrupted and pills that could have been drugs to regulate irregular cycles became the cause of irregular cycles for a lot of women. Without using the knowledge of the periodicity of the cycle to find the exact time of ovulation the cycle itself was disrupted so as to not have any ovulation at all. This disruption caused a lot of discomfort and pain but progress in contraception was seen less in alleviation of the pain and more in increasing efficiency by replacing the oral pill with other long acting hormonal methods that we discussed in the earlier section of this chapter. Efficiency and control have been crucial in this story of progress.

As an improvement over the hormonal contraceptives, the next idea that researchers worked with was that of anti-fertility vaccines. Today development of anti-fertility vaccines has not progressed along directions envisaged by the researchers in the 1980s and so we do not have an Anti-Fertility Vaccine (AFV) yet, but the idea of an AFV is something that reeks of the pharmaceutical industry and the medical profession's attitude to human bodies. As Dr. G. P. Talwar, the so called 'pioneering' scientist from India, said confidently on record to a documentary film maker even before the clinical trials had moved beyond the first phase, 'There will be no problem with anti-fertility (vaccine) because the contraceptive action is being achieved by chemicals that are organic to the body, which are being produced within the body' (Schaz 1991). To understand the arrogance and fallacy of this statement, it would be useful to understand how the AFVs work.

A vaccine, as we all know, is that which helps or initiates the process of generating an immune response through creation of relevant antibodies against antigens that actually cause disease. The situation is radically different in case of the AFVs. Here the antibodies have to be created against fertility or any of the proteins that are responsible for fertility. It could be the hormones which help production of egg and sperm, or those responsible for the implantation and growth or sustenance of the implanted embryo. There is no one causative factor for fertility and any of these multiple causes if stopped could prevent a conception from taking place. Hence, when one talks of an AFV, the first thing that has to be decided is the target antigen.

The next and bigger issue is of making the body believe that it needs to create antibodies against a chemical whose action it would assimilate as a natural mechanism. This virtually means that the body needs to be fooled so that it learns to create antibodies against substances which it would not normally react to. To be able to do this and to be able to ensure that the response is limited only to the chosen antigen, a part is selected from the total protein of the antigen which does not resemble any other molecule. This small fraction is then attached to a larger disease vaccine like diphtheria, tetanus toxoid or cholera. This modified vaccine is administered to the person whose body responds creating antibodies to the whole molecule.

For example, the trials carried out at the National Institute of Immunology, Delhi, selected as the target molecule the hormone human chorionic gonadotrophin (hCG) secreted in the woman's body after fertilization takes place. A small part of hCG, which does not resemble any other hormone structurally, was attached to a diphtheria or tetanus toxoid vaccine and this modified vaccine administered to the woman as an intramuscular injection. If an egg is fertilised in the woman's body, hCG would be secreted. But the antibodies created to act against a part of the hCG molecule would prevent that part of the hCG from functioning, thus neutralizing its overall function. Without the presence of hCG, the implantation cannot take place and thus contraception is achieved through the body's own immune system. This is what Dr. Talwar calls organically produced chemicals!

The arguments against the vaccine are manifold. Such a contraceptive method has no clear and distinct advantage over existing methods, which should be a requirement of all new methods. The vaccines researched have some inherent problems that cannot be resolved through more research; these are related to the concept of AVF itself. Most importantly, these can serve not as contraception but as methods of population control that can be widely abused and this apprehension is already proven in the ways that trials and research have been conducted. As one recent report says,

> The WHO, as a global coordinating body, has since the early 1970s continued the development of the Rockefeller-funded 'anti-fertility vaccine'. What also is becoming clear, is that extensive research has been done to the delivery systems in which these anti-fertility components can be buried, such as regular anti-viral vaccines. It's a mass-scale anti-fertilization programme with the aim of reducing the world's population: a dream long cherished by the global elite. (Maessen 2010)

This is the short history of contraception and modern medicine. More knowledge has meant more intervention in the processes of the body (only women's, of course) and less intervention in the process of sexual intercourse, more control in the hands of the doctors and an equal if not greater loss of control for the woman, more understanding of body processes for the researcher to make interventions and no new knowledge for the women in whose bodies these interventions will act, more power to the population control and market agendas and no real 'empowerment' of women or men. And yet these methods are used by many across the globe.

Needs Met, Yet Visions Unmet

The conundrum is that these technologies, however dangerous and harmful, do meet the needs of many women. Birth control is a long standing need expressed by women, something that ideally would help them explore their sexual desires with men in ways that are free and safe. Also even though the long term vision is of conversation between partners, responsibility in actions and respect for each other's autonomy, the present reality for most women is far from this at an individual level even in sanctioned socially approved relationships.

Contraception of any kind offers a feeling of safety from conception and thus opens up the possibility for subversions of social taboos and restraints–be they imposed by religion, community or even conservative political parties and organizations. In the church-dominated and anti-abortionist Americas and Europe in the 1960s, hormonal pills meant the 'sexual revolution' at least for some individual women–even though in the long run we may also see this as another patriarchal ruse, making some other women available against their wishes to male partners.

'Scientific' knowledge about the fertility cycle could in principle have provided another path which, however, has been very neglected. For the individual woman, knowledge of her own fertility is in itself an empowering process. Technological help to know the exact status of fertility can help alleviate some of the anxieties around conception whether one wants it or avoids it. This, of course, presumes the possibility of conversation and negotiation in sexual relationships. Women's groups across the country have shown that such knowledge empowerment does help build the process of communication. So in a sense the 'scientific' knowledge and the 'personal' experience of the physical body conjoin to create new empowering knowledges in consonance with feminist visions.

On the other hand, the biochemical invasive route that is largely taken is based on the notion that nothing can change in the dynamics of gendered societal power relationships. In this scenario it is difficult to imagine methods that involve the persons using them, that require some amount of alterations in the ways in which man-woman relationships and sexual desires are imagined, talked about and lived. My contention is that 'value-free' science actually reinforces multiple notions of the status quo and at times even contributes to further rigidity in ways that any bending and tweaking seems impossible.

And the technology that emerges from this value-free science adds its own complexity to the situation. It comes as a solution, a way out for the marginalized and the needy but actually reduces choices of various kinds. This is exemplified in many examples. The woman seeking to have some control over her reproduction, unable to negotiate with her male partner, not wanting a child as yet and not wanting the pain of another abortion, or the one who is compelled to take the option of contraception because rejecting it would mean less health care for her and her children, or the one who truly believes in the goodness of the medical system and its demi gods, the doctors, or the young woman who is not supposed to have sexual relations in the first place and accesses the available pill or injection as an easy solution, or the woman who has commercial sex and cannot negotiate condom use every time but does not want to get pregnant–for all of them maybe these technologies do seem like answers to their problems.

Similarly, an individual woman's need to be the 'good perfect' woman providing a 'healthy', 'normal', 'male' child to the patrilineal heteronormative family is met by the technology even if she pays with the cost of her health and well-being. And, of course, the cost that she chooses to pay will finally be weighed by her in her own context. But at a macro scale it leads to an overall disservice to the cause of women and the disabled.

Assisted Reproductive Technologies (ARTs) are another prime example where everyone in the local transaction appears to be benefiting. The doctor gets name, fame and, most importantly, business. The man has a child with a known confirmed genetic continuity, most often with his own genes. The woman produces a biological child for her family, hopefully of the 'right' kind either through giving her gametes, her uterus or both, or just by morally standing by the husband and his family and getting a child from someone else's gametes and uterus but treating it as her own biological child because the child has her husband's genes.

The donors of the gametes are financially compensated for donating either sperm or eggs. The male donor is still the one who has it easy because he needs no intervention in his body; he just gets an environment that the doctors think will help him masturbate and essentially is paid for masturbating. The woman donating the gamete donates what would have anyway been purged during her menstrual cycle. Of course, the extraction is not simple and there could be long-term effects but then again she operates from her understanding of what is safe. And finally if there is another woman whose body is used for the gestation of the child, she is probably doing it as reproductive labour and trying to get the best monetary compensation from it.

The question is who loses? In the long run the way in which this technology is regulated and implemented, it does the following: Instead of providing the possibility of questioning genetic continuity, it underlines the need for 'one's own child'; instead of questioning the need for a child per se and the socially constructed condition of pain of infertility, it underlines the pain and makes it essential to act on it; instead of making it possible for those rendered infertile by social norms like the single woman or the homosexual couple to have a child, the providers join hands with religious and social communities to make sure that they cannot access technology.

Finally, all those proposed individual solutions and local subversions are regulated and manipulated to create a more rigid world, which makes it difficult for anyone to make any other choice: The choice to not select any traits of the child, the right to not have a genetic child, the right to have a child-free life, the right to say no to the male partner whoever he may be–husband, new lover, a casual fling, the right to explore other ways of relating sexually, to expand the ambit of one's desire… essentially all that which is part of some of our feminist visions of our collective futures.

Feminist Agendas in Science: Reality or Fantasy

When as feminists or as women we ask for birth control or assistance in conception, what are we seeking? We are looking at the use of existing knowledge of our bodies to gain more control over ourselves. We want integration of our experiences and knowledges of our bodies with the formal body of knowledge called science to help negotiate difficult social terrains. Our innocent claim for value-free technologies is part of our idealistic belief that knowledge can help empower us in a way that some amount of negotiation is possible. We still do believe that there are ways in which these 'scientific' knowledges can evolve differently.

Knowledge production is a political subjective act and so the science of fertility that we seek as feminists will start with very many different questions and find more nuanced and different answers. It is not that as feminists we shall find that there is no hormone cycle that controls the periodic fertility cycle. But the hormone cycle may not have the same kind of prominence in our understanding of the human body; it may not be seen in this reductionist manner reduced to the concentrations of the hormones alone. The cycle has many dimensions and many of these might get attention.

How and why we take this understanding further will also be different. While the use of this knowledge and the applications we seek will definitely be different; so also will our concerns, our languages of description and analysis. We have a problem with this reductionist way of looking at human bodies or all life for that matter–applying the model of physics to all sciences as some would say. Recent work from biology and ecology in particular is challenging this in many ways and many other postcolonial and ecofeminists have voiced this discomfort with this approach of modern science. Understanding leading to control and intervention has been a trademark of modern science and technology and we hope that when perspectives change some of that would change too.

As Sandra Harding (2008: 120) says, 'No one's understanding can completely escape its historical moment; that was the positivist dream that standpoint approaches deny. All understanding is socially local, or situated. The success of standpoint research requires only a degree of freedom from the dominant understanding, not complete freedom from it.' We do believe in a feminist standpoint and agree with her, when she further adds (ibid.), that 'Women do not automatically have access to a standpoint of women or a feminist standpoint. Such a standpoint must be struggled for against the apparent realities made to appear natural and obvious by dominant institutions, and against the ongoing political disempowerment of oppressed groups.' Our standpoint does not grant us automatic access to greater truth or better knowledge. It just makes sure that where we come from is clear and not obfuscated under the guise of neutrality, as has been happening with science so far and in the scientific knowledge of reproduction and reproductive technologies for assisting or preventing conception in particular. What we arrive at is partial truth but the knowledge of it being partial itself makes it amenable to scrutiny and that is its strength.

The more complex question that we simultaneously need to address is the meanings of 'woman' that our feminisms are underlining. How do we strike the balance between the facts that women's bodies reproduce and that reproduction does not define women? What are the ways in which we deal with the contradictions of what women want and what we as feminists believe will help, that is, how do we nuance our differences around individual choice and the larger 'feminist' good? Are physicality and variations of desire included in our discourse around sexuality or are we sustaining a narrative of only love and emotion or violence and neglect?

Then there are questions around our somewhat 'pristine' feminist stands about the use of this knowledge. When we say no interventions at all, then are we also essentializing bodies? Is ability interwoven in our understandings of body, sexuality and reproduction? While saying that knowledge about the body helps empower 'women', how do we acknowledge the realities of those for whom the knowledge of being in a woman's body is itself a problem? Do our discussions around reproduction and fertility reinforce the body-based definitions of gender that are constantly being challenged by all those questioning the binary constructs of gender?

These and other such debates are at the core of the feminist visions of the body and the 'scientific' knowledges around it. This obviously cannot just happen by getting more women into the sciences. It needs feminist visions and understandings (which are thankfully always in the plural). And so it means constant evolutions of our feminisms and a scenario of debates and discussions to take all our concerns on board. Without an active feminist conversation it is impossible to think of any feminist science. The conversations have to be across the divide of academia and the movements, across the social and natural 'sciences', across disciplines. Transgressions are at the heart of it all.

Notes

1 It is interesting that this Nobel was announced on 10 April 2010, exactly 50 years after the contraceptive pill got FDA clearance on 9 May 1960.
2 It must, however, be clarified that the sexual revolution as openness in American society around sex and sexuality was already much underway before the pill was approved and marketed in 1960.

3   The subheading above is the title of a film made by Deepa Dhanraj who in turn got this phrase from a speech by the Indian prime minister Indira Gandhi who said that the population problem needed to be solved and in the process some people's human rights would be trampled because the situation is 'Something like a War'.
4   These interviews were conducted in 1991 with some women from Vadodara, Gujarat who had participated in the phase 3 multi-city clinical trials of Norplant in the 1980s.
5   The side effects that have been acknowledged as harmful and deserving of compensation include conditions like changes or irregularities in menstrual bleeding, pain, itching or infection at the implant site, swelling of tissue near the implant site, removal difficulties, headaches, nervousness, nausea, dizziness, breast pain, weight gain, and other such like. (Terri Davis et al. *versus* American Home Products Corporation, 94-11684 c/w (Civil District Court for the Parish of Orleans, State of Louisiana, USA, 22 March 2012).
6   The female condom has been a much later invention which works through creation of a covering along the whole vaginal passage. It is also interesting to note that diaphragms and cervical caps that were available through the Family Planning programmes vanished completely as soon as the other methods appeared and even in the cafeteria approach introduced later to apparently provide more choice to women, these have not made an appearance.

References

Allyn, David. 2001. *Make Love, Not War: The Sexual Revolution, an Unfettered History*. New York: Routledge. http://www.worldcat.org/wcpa/servlet/DCARead?standardNo=0415929423&standardNoType=1&excerpt=true. Excerpt accessed on 15 July 2012.

Balasubrahamanyan, Vimal. 1993. 'Fix It, Forget It: Norplant and Human Rights', *Economic and Political Weekly* 28, 22 (29 May): 1088.

Bandarage, Asoka. 1997. *Women, Population and Global Crisis: A Political-Economical Analysis*. London: Zed.

BWHBC, Boston Women's Health Book Collective. 2012. *History of Birth Control*. Retrieved 28 August 2012, from Our Bodies Ourselves: http://www.ourbodiesourselves.org/book/companion.asp?id=18&compID=53

Boston Women's Health Book Collective. 2011. *Our Bodies, Ourselves*. New York: Simon and Schuster. http://www.ourbodiesourselves.org/book/excerpt.asp?id=139. Excerpt accessed on 28 August 2011.

Chayanika, Swatija, Kamaxi. 1999. *We and Our Fertility: The Politics of Technological Intervention*. Mumbai: Comet Media Foundation.

Forum for Women's Health. 1993. 'Some Thoughts on Clinical Trials. *Medico Friend Circle Bulletin* 197–201 (August–December).

Ghai, Anita, and Rachana Johri. 2015, 'Science, Gender and Reproductive Technologies: A Case of Disability', in *Feminists and Science*, vol. 1, edited by Sumi Krishna and Gita Chadha, Kolkata: Stree.

Harding, Sandra. 2008. *Sciences from Below: Feminisms, Postcolonialities and Modernisms*. Durham, NC: Duke University Press.

Keller, Evelyn Fox. 1985. *Reflections on Gender and Science*. New Haven: Yale University Press.

Hartmann, Betsy. 1995. *Reproductive Rights and Wrongs: The Global Politics of Population Control*. Boston: South End Press.

INCITE website information on dangerous contraceptives http://www.incite-national.org/index.php?s=35

Infertility. 2012. Encyclopædia Britannica Online. Retrieved 28 August 2012, from http://www.britannica.com/EBchecked/topic/287593/infertility

Lepointe, B. 2010. 'Info Class Action: Depo Provera'. Retrieved 28 August 2012, from Recous Collective.info: http://www.recourscollectif.info/en/cases/depo-provera/

Maessen, Jurriaan. 2010. 'Rockefeller-Funded Anti-Fertility Vaccine Coordinated by WHO'. http://www.globalresearch.ca/index.php?context=va&aid=20906. Accessed on 28 August 2012.

Merchant, Carolyn. 1980. *The Death of Nature*. San Francisco: Harper & Row.

Nobel Prize 2010. 'The 2010 Nobel Prize in Physiology or Medicine - Press Release'. Nobel Prize Organization 11 October 2010. http://nobelprize.org/nobel_prizes/medicine/laureates/2010/press.html. Excerpt accessed 28 August 2012.

Rao, Mohan, and Sexton, Sarah, eds, 2010. *Market and Malthus: Population, Health and Gender in Neo-Liberal Times*. New Delhi: SAGE.

Saheli website. *https://sites.google.com/site/saheliorgsite/health/net-en* accessed on 20th May 2011.

Sarojini N. B. 2010. 'Women as Wombs'. *Infochange News & Features* (December, http://infochangeindia.org/agenda/medical-technology-ethics/women-as-wombs.html accessed on 18 September 2015.

Sathyamala, C. 2001. 'An Epidemiological Review of the Injectable Contraceptive, Depo Provera'. Medico Friend Circle and Forum for Women's Health.

Schaz, Ulrike. 1991. 'Antibodies against Pregnancy: The Dream of the Perfect Birth from the Laboratory'. *Documentary film. (English copy available through Ulrike Schaz, Hamburg.)*

Shiva, Vandana. 1988. 'Reductionist Science as Epistemological Violence', in *Science, Hegemony and Violence: A Requiem for Modernity,* edited by Ashis Nandy. New Delhi: Oxford University Press.

Silverman, E. 2011. 'Pfizer Settles Norplant Lawsuits for $29.5 Million.' (11 October). Retrieved 28 August 2012, from Pharmalot: www.pharmalot.com/2011/10/pfizer-settles-norplant-lawsuits-for-295-million

20

Feminism and Science: Present-Day Notes for a Feminist Standpoint Epistemology

Asha Achuthan

This chapter attempts to introduce, into the debate on gender and science, a question on experience as vantage point for knowledge-making. Indian feminists have already pointed to the lived experiences of 'women in the Third World' as resistance to modern western knowledge systems. I hope to examine and contribute to the debate by offering a different 'use' of lived experience—as *critique* as well as *resistance*. This use *connects* critique and knowledge-making, where *knowledge* is about comprehension of a possible world, and *critique* the aporetic comprehension of an actual world that makes this possible.[1]

I begin with two of my experiences in the field. These may be interpreted as different accounts of how hegemonic frames have accommodated anomalous experience. I go on to delineate the broad positions within feminism and gender work[2] that have responded to perceived knowledge-power nexuses, before returning

---

I would like to thank Gita Chadha, Sumi Krishna, Bindhulakshmi P. and students at TISS, for their insightful comments at various stages of this paper. I would also like to thank Raghavendra Gadagkar, Tejaswini Niranjana, Samita Sen, Ranjita Biswas, Nishant Shah, Nithin Manayath, Nitya Vasudevan and others who responded to the ideas at presentations of versions of this paper. I thank the Tagore Society for the opportunity to visit the Sundarbans and be part of some efforts. To the women of Sundarbans, I owe a *deeper* debt of learning.

to my field experiences in order to lay down the different 'use' I suggest above.[3]

Notes from a Consultation and a Conversation

The consultation

Tumi ki roj tablet khao? Do you have the pill every day?

Do *You* (the doctor and authority) have the pill every day?

Do you have to have the pill *every day*?

The conversation

In April 2002, I conducted, as a medical doctor, an intensive 6-day training programme organized by a non-governmental organization for 'traditional birth attendants'–*dais*–who had come from various parts of the Sundarban islands in West Bengal, India. The stated objective was to impart up-to-date, accurate scientific methods (adaptable to the field) of attending to pregnant women going into labour, in rural areas with limited access to hospital facilities. Local traditional practices could also be taken into account and legitimately incorporated *where useful* 'to fill in gaps in manpower at village levels', as the draft National Population Policy (NPP 2000) says. The dai, as yet uninitiated into 'method', in her own words the *mukkhu sukkhu manush* [an unlettered person], therefore is to have the key to a vast field of experience at births, a field waiting to be tapped usefully in development. Her know-how, which is 'practical' rather than 'propositional', means that she has no value in existing frames as epistemological agent; but hers is the voice of experience that with a degree of training and modification can apparently be made useful for the task at hand.

A few broad strokes about the form of the engagement are pertinent here, for they underline the apparent markers of difference between the (practically) experienced and the (theoretically) knowledgeable. Prior to introductions, the dais were asked to respond to a multiple-choice questionnaire highlighting problems generally faced during the delivery of a child. Later, the 'new', scientific methods were introduced and explained through lectures, models, role-playing, and video films (visual models had been deemed good methods, because a large number of the participants were non-literate in the conventional sense and unused to conventional methods of classroom learning). I would generally question them as to why they employed a particular practice, explain–in logical terms–why the scientific method was better, and then go on to demonstrate the functioning of the female body, as understood in (western) medical literature, with a ritual of endless repetitions, for the women were hardly used to the attention spans demanded of them. The 'students' enthusiastically took on the roles of woman in labour, dai, mother-in-law, husband, doctor at the local health centre, to enact the scenes as they should from now on be played out. The dai had come of age, as it were, in that 'classroom', and I had helped her up the ladder. Practices or understandings shared by the dais did afford me glimpses of knowledges that did not conform to (or sometimes compare with) the western episteme I was working with. But such difficulties I (had to) set aside for the purposes of my work. And following me, so did the dais.

The first question that the dais asked me when I arrived in their midst was whether I was married. If so, how many children I had. As I realized that I was alone in a room full of mothers, I felt the beginnings of an unbridgeable gap; I might pick up the local tongue, I might sit down with them and attempt to erase authority, but I did not share what they shared with most other women, the kind of experience they valued (or considered necessary for authority). As the classes wore on, this became a little joke amongst us–every now and then, one of the older women would stop proceedings to ask–*Accha, tomar to nei, tumi eto jano ki kore?* [How do you know, having none of your own?] And I would counter sagely–*Aro jani.* [I know that much and more] Finally they settled for–*Are eto rugi dekheche, ekta abhigyata hoy ni?* [She's seen so many patients, surely she must know something!] An experiential referent had been found, however clinical, and that was something!

For me, then, this was a scene of scientific as well as feminist empowerment. Seeing myself as the physician-feminist here, I concentrated largely on the gradients of power and resistance-negotiation operating between the dai, the 'development expert', the NGO, the local male quack doctor. The NGO had targets to meet–so many women over so many villages covered that year. The dais knew there was something in this for them. The kits that would be distributed at the end of session, the legitimation of their knowledge by the *sarkar* [government][4] – they were now trained dais, not just dais; this would hopefully carry meanings in trying times when the local (male) quack, armed with the 'injection'[5] and assorted other drugs, in short with a fair working knowledge of allopathic medicine, had all but edged them out of their already meagre income. I was doing 'research', and this was one of the ways I could listen in. Although I was there as the 'doctor', the authority, I saw myself, as feminist, altering the terms of engagement between these spaces.

The Turn to Experience: From Consultations to Conversations

I have no names (of protected confidentiality or otherwise) to offer for the women in both the episodes reported above; neither was part of an ethnographic study, and both are offered more as contexts within which feminist approaches to experience have materialized.

The *consultation* was at the family planning clinic of a state referral hospital, with a recalcitrant mother who had been put on the contraceptive pill following the abortion of an unplanned pregnancy and had returned for follow-up with a continuing carelessness regarding its intake. The entire consultation, as is evident from the report, lasted three sentences, leaving the female physician irritated, and the patient engaged in a certain conversational response–the kind that comes the way of the physician every day, but is nevertheless the kind that is illegitimate, aporetic.[6] Enough has been said about power-knowledge nexuses that promote 'one' knowledge–in this case the western medical–as high, as singular. It is clear that there are institutional and knowledge hierarchies at work here. This particular response is the kind that, through its own aporeticity–neither appropriate, nor oppositional, nor even alternative–makes visible, and bizarre, the claim of medical knowledge to objectivity and validity *on account of being unanchored to experience*. It is the kind of response that does not sit well with liberal feminist approaches that would wish to mediate authority through information, choice, or consent. This is also the response that plagues feminism in its present need to complicate simple notions of sisterhood, and we will return to it at the end of this chapter. Moreover, the somewhat bizarre turn this consultation takes is not entirely attributable to the apathy or non-personalized nature of care giving that is the feature of most large state hospitals, but more reflective of the detachment constitutive of the text of Science.

The *conversation* with the dai appears to be situated on the other end of the spectrum, not only free of violence and authoritarianism, but with a seeming acknowledgement of the experience and consequent value of the dai. My physician-feminist approach was working with this premise. On the one hand, I treated the hyphen physician-feminist as an uncomplicated and possible alliance; unaware of the infantilizations of the woman in Science, I presumed I was listening to the dai when I was actually indulging, allowing a little leeway before bringing her to heel. In the context of Science's responsibility to women, it seemed like a more inclusive approach that the feminist could introduce. At other times, I activated the disjoint in the hyphen, highlighting the knowledge differences. In the context of feminist critiques of Science, this was a response to the violence of orthodox Science. In the frame of the feminist turn to experience, it would seem to propose, through a rejection of Science's sole claim to expert knowledge, a first step towards recognition of women's knowledge. But is this indeed the case? What had I learnt from my practice with the hyphen?

In using the allegory of the two reports, we might see these possibilities of a faithful *turn* towards experience; I also wish to mark a possible alternative–that I call a *re-turn* to experience. I shall do this through a re-examination of the faithful turn; before that, I shall set up three examples of the turn that have marked feminist and gender work in the contexts of development–all of which would consider the classical consultation as problematic.

The Global Feminist Making of the Third World Woman: Building Capability, Fostering Agency

'Feminist political philosophy has frequently been sceptical of universal normative approaches. I shall argue that it is possible to describe a framework for such a feminist practice of philosophy that is strongly universalist, committed to cross-cultural norms of justice, equality, and rights, and at the same time sensitive to local particularity, and to the ways in which circumstances shape not only options but also beliefs and preferences' (Nussbaum 2000: 7). Nussbaum makes the case for feminist universalism in a world where universalism is more or less the declared enemy of feminism. Aware of the authoritarianism of the 'consultation' as symbolic of all mainstream systems, relating it to hierarchical sexual difference, convinced of the greater vulnerability of poor Third World women in this regard, she proposes the capability approach—a programme for building agency for this poor Third World woman who has been making do, while not having had the life conditions and choices to become 'uniquely human'—the biological and the pre-contextual subject of liberal philosophy. Nussbaum's version of the capabilities approach, proposed in basic agreement with Amartya Sen,[7] is a 'foundation for basic political principles that should underwrite constitutional guarantees' (ibid.: 70–71), and draws on 'Aristotle's ideas of human functioning. It is proposed as a universal and ethical approach that must nevertheless 'focus appropriately on women's lives' in order to be relevant, that is, it must 'examine real lives in their material and social settings'.

With respect to the consultation, the capability approach is available for a situation where the woman in the clinic is not indifferent, but aware of the conditions of her treatment, and able to make informed choices. An ethic of information is in place, as is consent. From the earlier restrictive knowledge motto that said *I know, you do*—visible in scientific as well as critical manifestos that worked with vanguard thinking, we moved to *we all know, together*. Authoritarianism has been undermined, perhaps . . . but what of authority? Nussbaum herself provides the answer when she says, in another context—'Why should we follow the local ideas, rather than the best ideas we can find?' (ibid.: 49). The logical connection being made here is the connection between expertise and authority, as above mere experience.

Is the female physician, then, within her rights, as an expert, to enjoin obedience, to accept that propositional knowledge is the way to go–'*Aamake niye katha hocche na?*' (It's not me we're talking about?) The constraints for capability that Nussbaum identifies are poverty, tradition, and lack of state commitments. The foundational assumption that allows her to propose universal strategies is that 'women's lives everywhere', and women's bodies, are essentially the same; where the seeming oddities are only differences in manifestation of stereotypes of women and men, rather than being signs of an 'alien consciousness' (ibid.: 23).[8] Women's lived experience, in other words, is the key.

Feminists and others within Third World situations stridently critique the construction of the 'Third World woman' that emerges in this and other western feminist scholarship through a discursive colonization 'of the material and historical heterogeneities of the lives of women in the Third World, thereby producing/ re-presenting a composite, singular "Third World woman"–an image which appears arbitrarily constructed, but nevertheless carries with it the authorizing signature of western humanist discourse' (Mohanty 1991: 53]).[9] This is a heterogeneity that may find common cause essentially only through struggle, not through biological or even geo-political or socio-cultural contexts. It is also a heterogeneity that is not taken into account in the 'victim of oppressive traditions' approach that such scholarship espouses.

But there is more to Nussbaum's case, and she responds to charges of cultural imperialism, ignorance of diversity, and paternalism, that are offered against universal values. A detailed discussion of these is beyond the scope of this chapter. But we return to the authority *versus* authoritarianism problem here. A feminist universalist like Nussbaum would wish to retain the framework of the consultation, and therefore the notion of the 'expert authority' and of 'one knowledge'. Indeed, she would see the indifference of the woman in that consultation as an absence of agency, as victimhood. She sees, through a building of capability, a situation where that woman can make the choice of contraception, or, if she refuses, can be 'forced' to do so in the interests of health and good citizenship as defined in international covenants. This poor, tradition-bound woman is 'typical' of the Third World, of the 'local' that asks for the exercise of a *non-imperialist universal* recognition of the particular *before it can be represented*, and this is what Nussbaum believes she is doing, when working with Indian women. And we cannot but agree with her in her posing of Third World women as a working category–plural and non-essentialist; any category, no matter how minutely contextualized, is by definition nominalist, unintended to capture the entirety of experiences, and to that extent, presence of heterogeneous experiences per se can hardly constitute a critique of category formation. Nussbaum's categories are, by her own admission, provisional, nominalist, *stable*, and hence not philosophically subject to the charge of rigidity. We will come back to this question with the debate on standpoint.

But what of authoritarianism? It is in the assumptions of the unimplicated, 'neutral outsider' that Nussbaum's universalism lies, as in her complete indifference to the anchoring 'sample populations' on which the ideal citizen, or the neutral definitions of reproductive health, for example, have been built. Herein lies the validity of Mohanty's charge of 'ethnocentric universality' (ibid.). A putting together of body-situation-circumstance that makes up 'Third Worldness' as a category of description for Nussbaum and her fellow-universalists, be it the embodied images of 'mothers of colour' breastfeeding their newborn, or the detailed physical descriptions of Vasanti and Jayamma–keywords of development discourse today[10] – and their surroundings, then, are not incidental to the narrative of their flourishing; it is, *singularly*, the *narrative* of the *particular*. In a frame of lack of capability, Vasanti or Jayamma can hardly be expected *not to have a body*; and they can hardly be expected to produce analytic statements. It is after this particularity has been described in its entire nuance that Nussbaum can set out to draw her comparisons with 'efforts common to women in many parts of the world' (2000: 22). And it is consistent with Nussbaum making her primary interlocutors these women, at the most mediated by Leela Gulati, the anthropologist in the field, not the feminists from the space who have attempted an interpretative account of the naming Third World.[11] It is here that 'experience' continues to retain valence as the necessary attribute of the Third World woman. And it is as an alternative possibility, of knowledge beginning from enmeshment rather than outsideness, that a speaking from experience is born–a rejection of consultations, a complete rejection of objectivity as a forgetting of experience, in other words: a soliloquy of the local: I know mine, you know yours, there is no dialogue.

The Third World woman as perspective *to speak from* has perhaps not been articulated as clearly anywhere else as in Maria Mies and Vandana Shiva's work on ecofeminism, work that speaks of the ways in which development becomes a powerful organizing metaphor for 'Third World feminism' (Shiva and Mies 1993). Building on the notions of organicity, wholeness, and connectedness, Mies and Shiva offer a picture of Third World women as already in convergence with nature, as upholders of the subsistence economy as against the 'capitalist patriarchal' system, and as offering perspectives for resistance to such an economy. Critiquing both western science and development, they endeavour to demonstrate the reductionist and universalist paradigms that the former occupies. For these critics, the 'mechanicity' that western science relies on, the ways in which it dominates nature-women–Third World, treating and re-producing each of these as a dead object, symptomatic of a subject-object dualism that is carried over into development philosophies too. These dualisms are seen as multiple and overlapping–Man/Woman, Reason/Emotion, Culture/Nature, Science/Experience–and ensure that a certain violence is constitutive of all scientific pursuits–evidenced in attitudes, institutional protocols, texts. It is this violence that this framework would see embedded in the consultation, and the response would be entirely different from Nussbaum's. Equality is not an option, as the woman's response is seen to come of radical *difference* rather than indifference. There is no place for this difference within western medical science, and therefore the only route is resistance, a rejection of the system. The consultation is violent; a conversation is impossible.

A Disaggregated (Third) World: Women Negotiating Meanings

But there is another kind of scholarship in currency that negotiates meanings of gender and women's lived experiences differently. This scholarship would reject consultation and the universalist approach, would recognize the asymmetry and hierarchies embedded therein, but would find resolutions in the conversation, identifying contingent moments of resistance, implicitly aware of *difference* and acknowledging experiential knowledge and a certain agency thereof, but *uninterested in questions of which knowledge might be valid*. This scholarship declares allegiance with postcolonial approaches in adopting a 'hybrid' approach. Anthropological investigations into midwifery and childbirth practices exemplify this position. This is also an approach that insists on the heterogeneity of experiences and the disaggregated nature of institutional apparatuses that apparently make a description of hegemony difficult,[12] and further, on the impossibility of even identifying such a hegemonic role for western science in the Indian context.[13] Alongside this disaggregation of biomedicine, this anthropology reads culture as 'in-the-making' through everyday practices also, thus immediately ranging itself opposite feminist scholarship that has pointed to medicalization of childbirth and pathologization of women's bodies and the presentation of the resistant indigenous as a solid, communitarian ideal. As Van Hollen (2003: 15) puts it, such disaggregations challenge 'those feminist studies that view all the controlling aspects of biomedicalized births as derived from a western historical legacy of the Enlightenment and the Industrial Revolution and that present a romanticized vision of holistic 'indigenous' birth, or 'ethno-obstetrics', as egalitarian, 'woman-centred' and non-interventionist'.

Van Hollen and others in this frame affirm feminist re-invocations of experience while needing to disavow existing feminist modes. Invoking the same historical and cultural specificity, and the heterogeneity of experiences of women's lives that feminism does, this gender work[14] nonetheless activates it to site resistance in the bricoleur-like response to various biomedical allopathic procedures rather than in a soliloquous 'natural therapy' movement. This, for Van Hollen, is the way out of 'the trap of representing others simply as victims' (ibid.: 10). This position, then, would represent modern medicine as weak and diversely articulated structures of power rather than a monolith. Women, in their negotiations, in their everyday resistances to this power, are agents making choices in context, but contingently so, not with the benefit of an ideological position like feminism.

Like much reflexive anthropology, Van Hollen performs the task of description with fidelity and often with ingenuity. This task of description is expected to offer a critique of macro-analyses, as also of rigid, monolithic description. In what often turns out to be a conflation of macro-analyses with generalization, of structural understandings with rigidity, but falls short. While representations of power as a disaggregated entity are meant to undermine rigid ideological stances and binaries, what gets reinforced is a narrative of Third World experience that may never consolidate into collective resistance or a universal change in perspective or a new epistemological position.

Alternatives? Notes on a Standpoint Epistemology

Both in feminism and gender work, therefore, there is consensus that the experiences of women's lives offer a perspective that is not accommodated in modern knowledge systems, and not reflected in knowledge-making. For some universalist feminists like Nussbaum, the consultation itself can be made friendly, for a recognition of the experience of oppression that Third World women pervasively suffer can make space for capability-building exercises for the victim, so that she too can reason, choose, set herself free.

There are several moves critical of the 'one knowledge' position. Those that take the Third World route either propose a 'different reason', a different canon, an alternative system (as postcolonial scholars sometimes did), or articulate a politics of complete heterogeneity that holds knowledge as necessarily provisional and separate from a rationale for politics (as did those that took on the name Third World feminism). A third position here is of *I know mine, you know yours, there can be no dialogue*, as mentioned earlier. For this position, the experience of oppression is necessary and sufficient. The consciousness of oppression, which is ex-officio, offers knowledge. The community of knowers here is a closed community. Asserting that the 'one knowledge' claim rests on the active exclusion of other knowledges, it suggests a remaking of 'low knowledge' through the *experience of oppression*. This impulse also allies with the organic or pastoral as opposed to the technological, an impulse sometimes tracing direct connections with a cultural past, and often offering a choice *between* systems of knowledge. The above mentioned Third-Worldist positions are sometimes tied up with this third position, proposing a politics of coalition while keeping knowledge bases separate (as in Third World feminisms), or proposing implicit connections between 'low knowledge' practices and a different system.

The politics of disaggregation alongside the complex of phenomena sometimes called globalization and the new vision of development as a practice, sometimes allies implicitly with this notion of cultural difference, but often explicitly rejects it. It rejects the counter-hegemonic effort as anachronistic, naming feminism as a narrow ideology that fails to respond fully to the heterogeneity of women's lives and experiences, and going instead with pragmatic choices and ways of knowing.

What, then, of alternatives? After a rejection of those feminist strands that seek to build a common, sometimes homogeneous *narrative of feminine experience*, after gender analysis that thrives on the heterogeneity of *women's experiences*, but yet agreeing with the need to 'speak from somewhere', as against older models of one knowledge that offered a 'view from nowhere', a neutral view, what could be the nature of this critique? If the consultation is obviously inadequate, and the conversation not enough, where might we look for a useful take on experience?

There are multiple methodological questions and pitfalls here. Where is somewhere? The 'somewhere' perspective would here mean something other than limited, or biased, as it has been labelled in 'one knowledge' frames. It would take on the third of three possible meanings,[15] as the fantastic spur within the dominant, as a moment of seeing, of 'possession', that can also be lost in the looking. Perspective as that moment of possession not only gives a completely different picture of things, it also gives a picture not available from anywhere else—that makes *visible* the dominant as such, as that which had rendered invalid other possibilities. This invalidation, this exclusion, could then be understood as a constitutive or primary exclusion with an entry later on the dominant's terms, and not as a removal from circulation of that which is disobedient, as commonsensical understandings of exclusion would have.[16] But how does this moment emerge, or from where? Here I find useful, as a beginning, the model of the excluded available within feminist standpoint theory (see also Chapter 14), of the woman as 'outsider within' who is favoured with this gift of vision on account of not being a full recipient of the system. While this formulation evokes a degree of unease about whether this social location can be enough as a starting point (whether women then always have to be the outsiders within to be able to speak from this space), it offers, I think, valuable clues for working toward a possible model of feminist critique.

Feminist standpoint theory talks of the possibility of a situated, perspectival form of knowing, of such a knowing as necessarily a communal project, and of this knowing as one where the community of knowers is necessarily shifting and overlapping with other communities. While Haraway (1992) would speak of 'situated knowledges' as against the 'God trick', as she calls it, of seeing from nowhere—a neutral perspective, Sandra Harding (1995; 2004) would go on to propose a version of strong objectivity—a less false rather than a more true view; this, Harding would suggest, can come only from the viewpoint of particular constituencies like women who have been traditionally marginalized or included only on certain terms. Starting off from these lives, from a recognition of social inequality, instead of a purportedly neutral position, is what maximizes objectivity, in her view. This is where Harding's version of standpoint epistemology is still grappling with the question of whether the experience of oppression is a necessary route to knowledge. (Harding deals with this with this by treating women's lives as resource to maximize objectivity; Haraway by treating these women as ironic subjects and seeing from below as only a visual tool.) A related question is whether the very notion of standpoint epistemology requires a version, albeit a more robust one than in place now, of systems of domination, and it is here that a productive dialogue could be begun between Haraway's more experimental version of 'seeing from below' and Harding's notion of strong objectivity. To Haraway's metaphors here–of the gift of vision, of the ironic subjectivity of those 'situated below the platforms of the powerful '(Haraway 1988: 583) – all of which point to the essentially contingent quality of the 'sighting' as well as the ability to see, I add the metaphor of possession—to flag that moment of seeing. In this sense, it is not possible to map perspective onto identity, 'pure' experience, or individual taste, since it is by nature contingent. (There is another point here, of whether such a moment can be inter-subjective, shared, or unique to that individual at that time, to which I will briefly return.) To understand this, we need to understand, also, that the issue here is not only that of recognizing and upsetting hierarchies. The very first example I gave, of the clinical consultation that turned into a conversation, tries to demonstrate this.

A feminist standpoint would be then the act of interpretation that puts this positioning, this transient possession, to work, not a place already defined. This 'somewhere', in other words, context is not (only) immutable date-time-place, but rather an inter-subjective quantity. It is most importantly about relationality, the space between you and me, both intra-community and inter-community. Once we take cognizance of this, we realize that that space does many things: it induces a porosity of boundaries (of body, community), it creates attachment, and also separation. We then have to talk of building a story from perspective, where it is the *turning from within outwards* (from attachment to separation) that does the work of building the story from that fantastic perspective, that moment of seeing. And such a standpoint 'is' only in the *constant interrogation* of both dominant discourse *and of the category of resistance* within which it may be named.

What we may have to gain from an attention to either consultations or conversations, then, is not so much the shift in form that we have made in moving from one to another, but the recognition of the fantastic perspective as a visual tool. Perspectives are made fantastic by their positioning in an imbrication of power *and* meaning; and unless the position is required to be static, they cannot be the source of a permanent identity, nor an alternative system. I present my report on the dai training programme, then, in a different detail and from a different perspective than as a look at indigenous systems of health or as a lesson to be learnt from women's experiences, or indeed as an essentially feminine perspective. I will attempt to delineate this in more detail now.

Returning to the Conversation

Let us go back to the dai training programme, mapping onto my narrative of it the paleonymies,[17] the weight of existing meanings, and possible difficulties of such a narrative. I have refrained from relating to this exercise as either participant observation (in anthropological mode) or as case study (the qualitative approach in medical parlance). Both of these, positioned at the same end of the methodological spectrum, were efforts that came up to serve a need for 'qualitative' analysis–the latter from within the scientific establishment, the former from within the social sciences. However, there is an effort to capture the microcosm that is a stepping away from earlier structural analyses; and a meshing of 'observer' and 'observed', a moving away from complete objectivity, that all self-respecting qualitative analyses undertake. These analyses are also an attempt to either expand or critique complete objectivity, and to take into account experience. This is what I have in mind when I refer to that time as 'conversation' rather than 'consultation'. What I am attempting here is a further *bracketing* of that effort, a bringing to bear, on the conversations, my identification of the problems with existing frames of critique.

This exercise will involve, therefore, a re-turn to experience. Such a re-turn will mean an attention to experience–not as narrative, resistant or otherwise, nor as fractured and unpredictable, but as aporetic–as affording a fantastic perspective on the dominant that had hitherto appeared as normal. An attention to the fantastic perspective will result in a turn from within (a community) outward–a different notion of the political from that of existing organizational, organic, or individual responses. It is, however, a notion that is hardly structural, a notion of the political as interpretation, but one that will have to be done each time. With these telegraphic steps in order, let us proceed. We had started the classes from the dais' voices—what they had written or what they had to say regarding their experiences with the births they had attended. The attendant presumption on both sides was that these voices were constituted by experience, the only prerogative of those uninitiated into *method*. I had then set about introducing a gentle reworking of the boundaries of this category 'experience'–till its quarrels with 'method' had diminished to negligible levels. Through the conversational mode, I had envisioned an ethical approach to 'one knowledge', through a minimization of violence.

How did I rework these boundaries? What were the terms of reference for the exchange between 'experience' and 'scientific method' that placed each firmly on a particular side of the divide between the untrained dai and the development expert, the body and the mind, the sensible and the transcendental? Several notions of the feminist political are in evidence here, working vis-à-vis dominant and other responses to the experience question. The responses may be charted in the following way. In the turn to experience as narrative, feminism has addressed the representation of the female body. The female body, we have seen, is the site for the understandings as well as operations of science (with its invisible qualifier western). In its project of defining the form and delineating the workings of the female body, scientific knowledge enjoys the status of a value-neutral objective method that purportedly bases itself on solid empirical evidence to produce impartial knowledge. In the case of the female body, it would then appear that science has *found* it exclusively and powerfully fashioned by *nature* to bear and nourish children; in the event, all it is doing is putting the facts before us.[18] Feminists have challenged the purported value-neutrality of science, making explicit the hidden cultural weight of scientific knowledge. For one, the standard referent body is male, by which the female body is judged small, inferior, or deviant; and through this a subtle process of othering or exclusion of the woman is instituted within science. Further, accounts of the workings of the body, its organs, its reproductive processes, are strewn with gendered metaphors that privilege the male as decisive, strong, productive, and the female, as complementarily passive, wasteful, unreasoning. It is against this authoritative, homogenizing strain that women's bodily experiences are posited in feminism—as something that is not only missed in Science's project of objectivity but something that is unable to articulate itself in and through Science's abstractions. In the event, the experience of the 'woman' within science is seen as that inassimilable, pre-discursive subjectivity that questions the *explanatory* potential of science, while also offering possibilities for agency.

There are certain collusions in the goals of these two projects, however, that bear looking at. Both are moving towards a single truth, whether derived from scientific theory or subjective experience, which they alone can represent. To this end, both homogenize and both declare the undisputed presence of this 'reality out there' that can be represented without mediations. And from here also flows a claim to objectivity. If Science posits a naturalized universal female body, experience would posit the 'woman' universalized through socialization. No experience can exist here outside the narrative, unless as aporia—the seemingly insoluble logical difficulty, much like the anomaly in Science. One would then derive that if scientific theories are built on exclusions, so is the category 'experience'. If Science claims value-neutrality, a simple valorization of experience ignores the 'historical processes that, through discourse, position subjects and produce their experience' (Scott 1991: 25). In the process, both Science and experience in turn achieve status as categories, homogeneous and uniform in themselves. Both discourses have the right to regulate entry.

If we then conclude that there is in this separation a certain essentializing of categories that ignores their very constitutions by the other, as also their constructions through cultural intelligibility, several questions arise. Can experience be that essential outside of Science that can grant agency? Or could it be also explicable as reflective of hegemonic norms that grant the sensible body as 'women's generic identity in the symbolic' (Whitford 1991: 136) while retaining a masculine topology for Science? This brings us to another feminist cognition of experience as constituted by history, circumstance, and as circumscribed by the norm as outside it.

Caught as I was between the conventional registers of science and feminism, I kept falling backwards into the question of results, and their reflection on validity. Experience, it would seem, was faulty by virtue of its very constitutivity, while Science continued to look rigorous and unbiased. As critical courier of scientific knowledge, I thought I was trying to weave myself into the discourse of the dais with minimum damage to their framework, and to that end I had decided to keep the question marks alive throughout, directing them towards science as well. But as I sat down to look at the assessment sheets on the afternoon of the first day's session, 'I' was fairly stunned. Of the ten questions put to the dais, one was worded as follows

> If the child does not cry soon after birth, we must– *(a)* say prayers over the baby; *(b)* perform mouth-to-mouth resuscitation; *(c)* rush the baby to the nearest health centre; (d) warm the placenta in a separate vessel.

Almost all 46 dais had affirmed the last answer. I remembered the asphyxiated babies that used to be rushed to the nursery in Medical College from the labour room that was on another floor. I remembered the bitter debates as to why the nursery was not stationed nearer the labour ward so that we could lose less time in resuscitation. I decided this could not be allowed to pass. And I conducted the classes accordingly. When we repeated the written examination at the end, none had ticked the last answer, and I was both relieved and vindicated. Later, I realized that I had succeeded only because I had adopted the authoritarian approach–right and wrong–to get across. And why had I done that? I realized, again, that with all my criticality, I was very much a scientific subject, and not merely because of my disciplinary training. I had retained reflexivity and criticality for as long as there was non-contradiction. Beyond that, I stayed put–well within Science. I too had my experiences–I could look at them as inseparably constituted by my production as scientific subject. I could imagine a different model of knowledge at work here that based itself on relatedness and not the atomism of the science I had been trained into. I could, should, have separated the question of this model from that of life and death. But I had been trained to look otherwise—at experience as non-knowledge, at knowledge as 'knowing that'—valid in every context, at the body as a series of events and items marked by their distinctness rather than their connections. And there I was.

In current development practice, though, the Third World woman is no longer considered to have no voice. On the contrary, she has a *specific* voice that is apparently being heard now in development projects in the Third World. In order to make this voice heard, however, she must have the capability to streamline it, make it universally understood as well as reasonable, and this is the cornerstone of the 'capabilities approach'. Here the dai, once named as dependable repository of traditional knowledge, can now be appropriated by notions of development flowing from liberal theories, for she also represents, in this frame, the rigid face of patriarchal traditions that have not given the woman voice. Development here is taken to mean empowerment—a granting, or rather restoration, of voice to the woman hitherto suffocated by tradition. The old order will indeed change, for the dais . . . *ek rakam chhilo . . . ebar anya rakam korte hobe* (things were different before, they will have to be done differently now) but that is hardly an exchange of tradition for modernity, or of experience for Science; it is an accommodation of one by the other. In the pluralism of current development discourse, the dai is a figure who exists before context, occupies an underprivileged class position, and has a voice that may be streamlined into the mainstream.

And in feminism, despite, or after, the recognition of 'women's experience' as constitutive of hegemonic norms, there is a renewed positing of experience as resistant, as the natural habitat, perhaps, of the woman. This is, of course, clearly in evidence in what I have called the global feminist undertaking, which is best argued philosophically in Nussbaum's work, and most tellingly represented in her examination and insertion of 'Jayamma-the-brick-kiln-worker'—who *cannot not* have a body that speaks—into the lexicon of development literature. As Third World women's practices that contribute to culture-in-the-making, it is visible in the gender work that I have talked about.

What of my 'conversations' with the dai? As medical-professional-feminist addressing-gendered-subaltern, I recognized and tried to steer clear of the various precipitations of such a binary; I ended, however, looking for a connection *through experience* between the 'professional' and the 'unlearned'; for an essence to the feminine, perhaps, or to woman in the Symbolic. The earlier legacy of experience inheres here; in asking questions of an epistemic status for experience, in the anxiety of not being able to accord it equal validity, in looking for a separation between feminist critical projects and dominant discourse through a recourse to a feminine difference which will be different from the place accorded to women in the patriarchal Symbolic.[19] Most telling, perhaps, it inheres in the anxiety over the similarity or otherwise of perspective between the (feminist) professional and the (woman) dai; one that presumed that the origins of an organic connectedness was to be found in the unspoilt dai who talked of *meyeder meyeder katha* [this is between us women].[20] So the first attempt that the dais made to connect with me was through *abhigyata*–experience. And the overwhelming feeling at the end of those six days amongst the dais, and in me, was of a solidarity that had perhaps been established across boundaries of authority (though not disruptive of it in any way), across science, across different experiences. But . . . is this how feminist projects are going to differ from development initiatives? Does the Third World woman indeed exist? While accessing a connectedness that would not mean the place accorded to women in the patriarchal Symbolic would definitely be a move, where would this connectedness be situated? If not in family or traditional community, would it be in some other sense of being together? Will we seek to continue its residence in women? Will we travel from an erasure of experience, the feminine, the subjective, to an essentializing of the same? Will women be the 'embodied others, who are not allowed *not* to have a body, a finite point of view'? (Haraway 1988: 169). If so, are we still going to stay with the biological body as pre-discursive resource of experience? If science is to remain the ultimate arbiter, is experiential agency then to be only the aporia, showing up as resistances through gaps in policy? Or can there be a feminist policy-framing that can work on the aporeticity of experience?

It is here that ideas put forth by feminist epistemologists–on gender experiential ways of knowing, on knowing how instead of knowing that (Dalmiya and Alcoff 1993), on empathetic connections, on responsible listening (Code 1988), and on care-knowing (Dalmiya 2002), might provide notes towards a way through the thickets of the very category 'women's experience'. There might perhaps be a *meyeder meyeder katha* that keeps the dai and the woman in labour closer together than the female obstetrician and the woman in labour, simply on account of a *recognition* rather than *rejection* of the commonality of their experiences, and an empathic, attached model of knowledge that acknowledges this. Such a model of knowledge could well be a 'knowing how' that cannot be 'said', or codified in terms of the propositional 'knowing that', as also one that incorporates an ethics of care and relatedness, in direct contradiction to the detachment necessary for traditional propositional knowledge (Gilligan 1982; 1995). (Gilligan has responded extensively to the dangers of such a connection of care for women, proposing a clear difference between a feminine ethic of care that is self-undermining and a feminist ethic of care that is empowering, shared, and not essentially attached only to women, although perhaps a model drawn from women's lives). For the fe(male) obstetrician, there might be a chance of learning to share this space, but perhaps without the edge that the first connection might carry–again on account of the propositional models that engender the consultation and modern obstetrics. The sharing of that space will involve, then, acts of responsible listening (Code 1988), and interpretation that utilize the visual tool we spoke of earlier, that gift of vision (Haraway 1988) available to the outsider within.

I pick up, therefore, on the gaps in the quintessentially anthropological narrative, to bring back the question of aporeticity. We have spoken extensively, in feminism, of the fractured narrative. Rather than the fractured *narrative*, however, it might be the *fracture* we need to speak of now. And rather than look at women as being essentially capable of *mimetisme*,[21] and therefore as the essential content of fracture, it might be useful to access the *moment* of fracture, using as allegory, not narrative resource, the responses of the dais to the reproductive health apparatus, or the bizarre consultation between the recalcitrant mother and the female physician. Let us not hunt for simple, organic connections between me (trained elsewhere) and the dai as women as my resource, but acknowledge our very asymmetry of dialogue, our seeming separation. Such a concentration on momentary fractures, disallowing as it does a final and fixed concentration on 'woman', or a continuing separation of registers between politics and knowledge on account of the 'fantastic' perspective opening up a fresh vantage point both of knowing and critique of possible worlds, I submit, would constitute newer notes towards a feminist standpoint epistemology.

Notes

1 I use the word 'aporetic' here and throughout the chapter to speak of the logically insoluble theoretical difficulty, the impasse. The French word *aporie* is derived from the Greek *aporia*, meaning difficulty, that which is impassable, especially 'a radical contradiction in the import of a text or theory that is seen in deconstruction as inevitable' (*Merriam-Webster Dictionary Online*).
2 I use the term as a catch-all to refer to a lot of research appearing particularly in the 1980s and 1990s that pertains to women in particular, that distances itself from feminism. A detailed explanation follows in the later part of the paper.
3 The consultation comprised the three sentences; the indented paragraphs that carry the seemingly bizarre statements uncover the underlying assumptions in the scientific text that we will come back to in the next section.
4 Sarkar means government. For the dai, however, the analytic separation between government and non-governmental organization is a blurred one. The space of civil society that the NGO conceptually occupies as separate from the state is unavailable to her; both represent the call of legitimate authority that have brought her here. And yet, does her turn to authority have an element of the conscious? Poltidi (name changed, Polti Jana, one of the economically more disadvantaged

Feminism and Science 169

of the group, also one of the most attentive and eager to imbibe the new) approached me the day after the video film showing a trained dai at work in Rajasthan. She had watched the dai in the film fill up her register with the details of each birth she attended, and report to the municipal office, and had come with a request for us to arrange something similar for this group. So that, as she understood, they could make an honest (and just) living, for in such a case payment to the dai would presumably be fixed and commensurate to her efforts.

5 Oxytocin, used (under strict monitoring in hospital settings) to induce uterine contractions, and used freely by these practitioners when called in to assist at delayed labour, with effects ranging from the magical to the disastrous.

6 For each part of the consultation, I have set down, in the indented paragraphs, those unspoken, seemingly bizarre statements that uncover the philosophy of the propositional model of knowledge at work here, and that consequently determine the consultation as valid.

7 Amartya Sen is known for his contributions to welfare economics, particularly human development theory, understanding the underlying mechanisms of poverty, gender inequality, and political liberalism. He has addressed problems related to individual rights, justice and equity, majority rule, and the availability of information about individual conditions.

8 'The body that gets beaten is in a sense the same all over the world, concrete though the circumstances of domestic violence are in each society' (23).

9 Although the arguments quoted here are from Chandra Talpade Mohanty's text (1991) published well before Nussbaum's, and although Mohanty's critique is specifically based on the Zed Press's 'Women in the Third World' series of publications (as being 'the only contemporary series . . . which assumes that "women in the third world" are a legitimate and separate subject of study and research' [75, endnote 5]), Nussbaum has already been expressing her position vis-à-vis the capabilities question from the 1990s itself, drawing on Aristotle as a resource for an account of human functioning. Further, Mohanty's work seems to read directly, critically, and powerfully into some of the concerns in Nussbaum's self-avowed feminist political philosophy, particularly her writing on women in the Third World that largely follows the

women-in-development approach. Mohanty has been one of the more vociferous and visible critiques of first world feminism, and as such, it is necessary to engage her critique at this point. There are also significant ways in which Nussbaum's text shows up shifts in thinking in First World feminisms themselves, and it is with these in mind that I juxtapose the two.

10 Breastfeeding is part of the exercise of Third-Worlding that was promoted by development agendas and globalist feminist rhetoric alike. It was promoted, ideologically and pragmatically, as the battle against bottle and artificial feeds, as the alternative to global Capital, thus making the Third World mother a self-sufficient provider of nutrition, and as the metaphor of responsible motherhood–as the embodied, as the reservoir of feminine experience and resistance. See http://www.unicefiec.org/document/poster-on-exclusive-breastfeeding-for-six-months. Accessed on 4 November 2015.

11 Stories of 'two women trying to flourish' as perceived and told by Martha Nussbaum. 'Unlike Vasanti, Jayamma has been examined previously in the development economics literature . . . I am very grateful to Leela Gulati for introducing me to Jayamma and her family and for translating' (Nussbaum 2000: 17, fn. 21). Leela Gulati, known for having brought anthropological perspectives to bear for the first time on seemingly economic issues, was the first to discuss widow and brick-kiln worker Jayamma in her work on widows in India, in Martha A. Chen, ed, *Widows in India: Social Neglect and Public Action* (New Delhi: SAGE, 1998), and also in other work on women's studies perspectives.

12 'Anthropologists have begun to examine the diverse and uneven ways … [in which] childbirth is being biomedicalized throughout the world' (Van Hollen 2003: 15).

13 'Unlike the situation in the United States and many parts of Europe, the biomedical establishment's control over childbirth in India can by no means be viewed as hegemonic' (ibid.: 55).

14 I use the term as a catch-all to refer to research appearing particularly in the 1980s and 1990s pertaining to women in particular, which distances itself from feminism. A lot of this work is to be seen in development scenarios. Gender work reflects the change from the more obvious ideological critiques of development to post-development positions. From earlier feminist critiques suggesting that 'woman', or the 'woman's body',

has been the grounds of colonialist discourse, and drawing continuities between this and development language, these arguments point to both the absolute heterogeneity of the experiences of women in the Third World and the active contribution of these experiences in constituting a response to development policies. This gender work, largely hosted in anthropology and happening in the 1990s, has taken on the task of describing the ambivalent relationship that women's lived experiences have had with Science as an institution, and so with development as underwritten by Science; they have taken up issues of context or location, personal experience, and marginalization, at each point attempting to reverse the devaluing of women's experiences. The attempts have ranged from rationales for inclusion in development programmes, to descriptions of negotiation, to declarations of independence from development logic. It might also be noted that the word 'gender' in this space is a descriptive use; as against the use of gender as a category of analysis that was one of the triumphs of feminism in its movement from 'women' to 'gender'.

15 Three meanings of the word 'perspective' provided by the Oxford English Dictionary: 1. The relation or proportion in which the parts of a subject are viewed by the mind; the aspect of a matter or object of thought, as perceived from a particular mental 'point of view' … [h]ence the point of view itself; a way of regarding (something). 2. A picture so contrived as seemingly to enlarge or extend the actual space, as in a stage scene, or to give the effect of distance. 3. A picture or figure constructed so as to produce some fantastic effect; e.g. appearing distorted or confused except from one particular point of view, or presenting totally different aspects from different points.

16 'Exclusion in principle seems to function as a formidable method of forcing dependence. And it is indeed a choice between 'being on the outside or perhaps at my heel', conveying first an exclusion in principle, and then conditions for secondary entry, rather than the reverse, 'at my heel or on the outside', which would indicate first a frank authoritarianism and then punishment for insubordination' (Le Doueff 2003: 25).

17 The word 'paleonymy', as suggesting a history of earlier meanings for a word, or practice, is best described by Spivak in conjunction with the idea of language always also not being an adequate fit, thus: 'Whenever someone attempts to put together a "theory of practice" where the intending subject as absolute ground is put into question, catachrestical [words that have been used wrongly; ie they mean something else]

master-words become necessary, because language can never fully bypass the presupposition of such a ground. "Value," and consequently "value-form," are such words for Marx. The particular word is, in such a case, the best that will serve but also, and necessarily, a misfit. ... The choice of these master-words obliges the taking on of the burden of the history of the meanings of the word in the language (paleonymy). Thus "value" (as "writing" in Derrida, "power" in Foucault, or yet "desire" in Deleuze and Guattari) must necessarily also mean its "ordinary" language meanings, material worth as well as idealist values, and create the productive confusion that can, alone, give rise to practice.' (Spivak 2012: 334).

18 This would be stressing the empirical foundations of Science, but Human Sciences have always been the area where the subjective is most easily detected–hence the name 'soft sciences'. Things are changing, however, with the Biological Sciences rooting themselves in the 'knowable' gene–their accession to hard objectivity is now a reality.

19 The place of women–in patriarchy, in a language outside patriarchy, has been a recurrent theme in the thought of Luce Irigaray. Interpreting Plato's myth, she draws a picture of the analogies with the patriarchal arrangement, and proposes another topology. Plato's Idea she designates as the realm of the Same–'the hom(m)osexual economy of men, in which women are simply objects of exchange . . . The world is described as the 'other of the same', i.e., otherness, but . . . more or less adequate copy but . . . woman is the material substratum for men's theories, their language, and their transactions but . . . the 'other of the same' but . . . [or] women in patriarchy but . . . the 'other of the other' but . . . is an as yet non-existent female homosexual economy, women-amongst-themselves but . . . In so far as she exists already, woman as the 'other of the other' exists in the interstices of the realm of the [Same]. Her accession to language, to the imaginary and symbolic processes of culture and society, is the condition for the coming-to-be of sexual difference.' See 'The same, the semblance, and the other' in Whitford (1991: 104).

20 This is a common saying in Bengali that carries connotations both of an exclusivity–a woman's domain–as well as insignificance–this is just something between us women.

21 To travel from 'mimesis imposed' (Irigaray's term for the mimesis imposed on woman as mirror of the phallic model) to 'mimetisme'–'an act of deliberate submission to phallic-symbolic categories in order

to expose them', where 'to play with mimesis is ... to try to recover the place of but ... exploitation by discourse, without but ... simply [being] reduced to it ... to resubmit but ... so as to make "visible", by an effect of playful repetition [mimicry, mimetisme] what was supposed to remain invisible ... 'is the Irigarayan project' (Irigaray 1991, quoted in Diamond 1997: 173).

REFERENCES

Code, Lorraine. 1988. 'Experience, Knowledge and Responsibility'. In *Feminist Perspectives in Philosophy*, edited by Morwenna Griffiths and Margaret Whitford. Bloomington, IN: Indiana University Press, 1988: 187–204.

Dalmiya, Vrinda. 2002. 'Why Should a Knower Care?' *Hypatia* 17, 1: 34–52.

Dalmiya, Vrinda, and Linda Alcoff. 1993. 'Are "Old Wives' Tales" Justified?' In *Feminist Epistemologies*, edited by Linda Alcoff and Elizabeth Potter, New York: Routledge.

Diamond, Elin. 1997. *Unmaking Mimesis: Essays on Feminism and Theater*. London and New York: Routledge.

Gilligan, Carol. 1995. 'Hearing the Difference: Theorizing Connection', *Hypatia* 10, 2 (Spring): 120–27.

———. 1982. *In a Different Voice: Psychological Theory and Women's Development*. Cambridge, MA: Harvard University Press.

Haraway, Donna. 1992. *The Promises of Monsters: A Regenerative Politics for Inappropriate/d Others*. In *Cultural Studies*, edited by Lawrence Grossberg, Cary Nelson, Paula A. Treichler. New York: Routledge.

———. 1988. 'Situated Knowledges: The Science Question in Feminism and the Privilege of Partial Perspective', *Feminist Studies* 14, 3 (Autumn): 575–99.

Harding, Sandra, ed. 2004. *The Feminist Standpoint Theory Reader: Intellectual and Political Controversies*. New York: Routledge.

———. 1995. '"Strong Objectivity": A Response to the New Objectivity Question' *Synthese* 104, 3: 331–49.

Le Doeuff, Michelle. 2003. *The Sex of Knowing*. Trans. K. Hamer and L. Code. New York: Routledge.

Mohanty, Chandra T., A. Russo, L. Torres, eds. 1991. *Third World Women and the Politics of Feminism*. Bloomington: Indiana University Press.
National Population Policy (NPP) 2000. http://mohfw.nic.in/showfile.php?lid=2949; last accessed on 11 November 2015.
Nussbaum, Martha C. 2000. *Women and Human Development: The Capabilities Approach*. New Delhi: Kali for Women.
Scott, Joan W. 1991. 'The Evidence of Experience', *Critical Inquiry*: 773–97.
Shiva, Vandana, and M. Mies. 1993. *Ecofeminism*. Halifax: Fernwood Publications.
Spivak, Gayatri Chakravorty. 2012. *Outside in the Teaching Machine*. London, New York: Routledge.
Van Hollen, C. 2003. *Birth on the Threshold: Childbirth and Modernity in South India*. New Delhi: Zubaan.
Whitford, Margaret. 1991. *Luce Irigaray: Philosophy in the Feminine*. London: Routledge.

ADDITIONAL READINGS

Bordo, S. 1994. 'Feminism, Postmodernism and Gender Skepticism'. In *Theorizing Feminism*, edited by A. Herrmann and A. Stewart. Boulder, Co: Westview Press: 458–81.
Butler, J. 1993. *Bodies That Matter: On the Discursive Limits of 'Sex'*. New York: Routledge.
Chaudhury, A. K. 1987. 'In Search of a Subaltern Lenin'. In *Subaltern Studies V: Writings on South Asian History and Society*, edited by Ranajit Guha. New Delhi: Oxford University Press.
Foucault, M. 1970. *The Order of Things: An Archaeology of the Human Sciences*. New York: Vintage.

21

Science and the Making of a New Nationalist Masculinity in Colonial Bengal

Madhumita Mazumdar

The feminist critique on science has not only concerned itself with equity studies aiming to understand the under-representation of women in science but has also engaged with informal mechanisms of discrimination against women that derive from larger sexist, racist and classist projects, which have used science for support and legitimacy. To decode the political underpinnings of such projects the focus of feminist critique has had to move from the more obvious institutional sites of scientific practice to larger social and cultural domains where science in general has colluded in the historical perpetuation of gender inequities (Harding: 1991). Feminist scholars over several decades have also argued that structural inequities of gender in science require persistent historical explanation over time and across cultures.[1] Such historical studies, for instance, can help explain why after decades of women's participation in science in India the popular imagination of science and scientist remains deeply imbricated in masculinist ideologies.

This chapter draws its arguments from this strand of feminist critique and tries to understand the mutual constitution of science and the patriarchal projects of anti-colonial nationalism in Bengal. It seeks to explore ways in which science and scientists entered nationalist discourses on masculinity that sought to counter colonial stereotypes of Bengali men, with the elaboration of a new masculine type that was both recognizably 'modern' and 'national'.[2] It tries to develop the argument that imagining the scientist as masculine in a quintessentially 'Indian' way, implicated scientists and scientific practice in what has been described as 'symbolic gender'–'a central organizing discourse of culture, that inflects not only ways in which men and women experience and understand themselves as such but also interweaves with other discourses and shapes them' (Cohn 1993: 228).

Apart from unravelling the series of complex cultural negotiations that marked the shaping of this project, the larger purport of this chapter is to understand the historical underpinnings of the enduring perception of science as a 'one-gendered world'. Here I draw attention to the strand of historical literature on early modern science that has looked into the several ways in which science and scientific practice came to be associated with discourses of masculine self-fashioning and self-care.[3] Jan Golinski in his survey of the writings of Francis Bacon, René Descartes, Robert Boyle, John Locke and Isaac Newton develops the argument that the masculinist imaginings of early modern science may not have originated only in the overtly misogynistic assumptions of Bacon's writings with their demeaning characterizations of women and their domineering attitude towards nature as had been hitherto argued by Merchant and Keller (Merchant 1980; Keller 1985). Golinski (2002: 127) following what he describes as an 'alternative' reading of Bacon, argues that rather than advocating the seduction and rape of nature as commonly perceived in early feminist literature, Bacon proposed the scientific inquirer to devoutly submit himself to a divine revelation of truth.[4] In other words, Bacon argued that the domination of nature demanded preparatory exercises of self-purification and discipline. It was this idea of scientific pursuit with its insistence on self-mastery through bodily regimes, mental drills, the control of passion and spiritual self-examination as a condition for the acquisition of reason that implicated the Baconian science in a profoundly masculine ethic.

Though far removed from seventeenth-century Europe, the Baconian ethic of scientific practice and the discourses of masculine self-mastery that it propagated, had strong resonances in nineteenth-century Bengal. The almost seamless interweaving of this Baconian discourse with that of the cultural nationalist imaginaries of science in the late nineteenth and early twentieth centuries constitutes the primary focus of this chapter. It takes up for discussion a selection of cultural artefacts and discourses that demonstrate the critical reconfigurations of the Baconian project in cultural nationalist discourses of science that took shape in the years between 1839–1931. It explores the common discursive thread that linked Pandit Iswarchandra Vidyasagar's didactic biographies of scientists to Rabindranath Tagore's poetic eulogies of Jagadish Bose, the architectural symbolism of the Bose Research Institute, the autobiography of Acharya P.C. Ray and the popular biographies of scientists penned by Anil Chandra Ghosh. It tries to show how the discourse on character and improvement in Vidyasagar's *Lives of Scientists* was reworked in the nationalist imaginations of Tagore, Bose and Ray that reinforced the Baconian ethic of science as a pursuit of masculine self-discipline and discovery but steered it away from its individualistic and utilitarian objectives. They embedded it instead within a patriotic discourse of liberation of the colonized motherland and the reclamation of national glory. In this reworking of the Baconian project I argue, the links between scientific rationality and masculinity was both 'nationalized' and 'naturalized' in ways that effectively pushed Indian women scientists outside the frames of public visibility.

Although the focus of this chapter is on the discursive framing of nationalist science and masculinity it remains attentive to the fact that such framings were deeply embedded in the structural–material conditions of scientific practice for Indians in the late nineteenth and early twentieth centuries. The facts of limited colonial patronage, combined with structural inequalities in academic institutions and racial prejudices associated with colonial forms of rule have been well documented by historians writing about the making of the Indian scientific community in colonial Bengal.[5] Yet what has seldom been commented upon are the implications of this on the emerging trajectory of indigenous patronage of scientists and scientific practice in the late nineteenth and twentieth centuries. I foreground the issue of indigenous patronage to make the point that the early links between nationalism and the gendered discourses of scientific practice were forged within the complex contours of science patronage. While most historical accounts concerned with the growth and development of a *national* science in Bengal locate their point of entry in 1876, the year of the establishment of the Indian Association for the Cultivation of Science by Dr. Mahendra Lal Sircar, this study begins the story on a slightly later date.[6] This is not to devalue the initiative of Dr. Sircar, in setting up the basis for 'national science' inside the portals of the Association, but to look at the story of 'national science' from the moment of its first *projections* as such.

Starting from this position, this chapter tries to develop the argument that the beginnings of *national* science in colonial Bengal needs to be seen not merely through instances of indigenous patronage and participation in science, but in attempts to define the distinctiveness of the 'Indian' endeavour.[7] The imperative of drawing public support and patronage for Indian science in other words demanded a new cultural imaginary of science; one that was closely affiliated to the nationalist political endeavour. When seen from this perspective, it becomes possible to chart the range of cultural initiatives, which mediated the links between science and the nationalist agenda in colonial India. The glimmerings of one such initiative, which worked to define the cultural distinctiveness of the Indian scientific endeavour, was evident in Rabindranath Tagore's interpretation of the aspirations and achievements of Bengal's first physicist, Acharya Jagadish Chandra Bose. The terms in which the poet projected the attainments of Bose, in a sense set the tone for a new nationalist engagement with science in colonial Bengal.

To explore the shifting equations between science, nationalism and masculinity, I draw attention to Gyan Prakash's *Another Reason: Science and the Imagination of Modern India* (2000), which suggests that colonial Bengal saw the making of one of the most powerful creative projects around science. The engagement with science Prakash argues, which began as part of a liberal social reformist agenda in the mid-nineteenth century, was gradually subsumed within an emerging wave of cultural nationalism in the late 1890s. The ideological thrust of this project was of critical import to the larger nationalist agenda, of fashioning a culture that was both modern and recognizably Indian.[8]

Science and the Making of a New Nationalist Masculinity 179

Partha Chatterjee in one of his early studies of nationalist discourse (1993) had argued that the political rationale of anti-colonial nationalism in Bengal limited the possibilities of any fundamental cultural negotiation with either the epistemological or the ethical intent of modern science.[9] This was largely because of western science's purported *location* in the scheme of nationalist political action. He later elaborated this point to show how in terms of its internal logic, anti-colonial nationalism in Bengal created its domain of sovereignty, by dividing the world of social institutions and practices into two identifiable domains: the 'material' and the 'spiritual'. The 'material' being, the domain of the 'outside', of the economy, and of statecraft, of science and technology, in sum, a domain where the West had proved it superiority and the East had succumbed. The 'spiritual', on the other hand, was marked as an 'inner domain' bearing the essential marks of cultural identity. The greater one's success in imitating western skills in the material domain, the greater was found to be the need to preserve the distinctiveness of one's spiritual or cultural domain. It was thus within the 'inner domain' of national culture, that nationalism launched its powerful and historically significant project to fashion a modern national culture that was nevertheless not western. He drew attention to the ways in which nationalists in Bengal sought to create a new 'national modernity' around the institutions of language, religion, literature, art, education and the family. Science, in this cultural project, remained conspicuous by its absence.[10]

Prakash (2000) sought to revise this formulation by showing how the impulse to inscribe onto western science the markers of Indian cultural distinctiveness was a trend which began in the late nineteenth century and culminated finally in the later nationalist initiatives to include science as the ideological prop of the independent Indian nation-state. What Prakash failed to emphasize, however, was that the exigencies of this nationalist project demanded multiple negotiations over time and with the multiple forms in which science as an institution of colonial modernity came to inhabit the realms of state, civil society and the 'inner domains' of national culture as described as Chatterjee. For it is in these 'inner domains' of national culture including language, art, education and family that science came to be embedded in the gendered discourses of nationalism.

Though seldom stated explicitly in these terms, both scientists and their patrons and interlocutors in colonial Bengal drew upon the political and cultural discourses of nationalist/Swadeshi to define the ethics of scientific practice and represent the persona of the scientist as the heroic bearer of a distinctively nationalist masculinity. Critical to this formulation was a definition of scientific practice as a heroic pursuit that demanded the spiritual and ethical disposition of the *practical ascetic* and the political sensibilities of the true Swadeshi.[11] Both categories were deeply embedded in a gendered discourse of masculinity that had begun to take shape through a range of political and cultural writings of the period.

I begin by briefly introducing the early conventions of scientific biography writing in colonial Bengal popularized by the Bengali literatteur and social reformer Pandit Iswarchandra Vidyasagar. I try to show how these early conventions of scientific hagiographies/biographies combined Victorian and brahminical discourses of 'character' to figure the scientist as an icon of exemplary masculinity. This is followed by an account of the role of the poet Rabindranath Tagore who as friend and patron of the scientist Jagadish Chandra Bose intervened to shape the emerging discourse of science and masculinity in new and unexpected ways. Though inspired by the earlier discourses of Hindu masculinity, Tagore's idealised image of the scientist as *rishi*, a sage, was meant to transcend the bounds of religion or ethnicity and elevate scientific practice from the domains of the everyday to the realms of spiritual contemplation. Tagore's intervention it is argued, marked the constitutive moment in the formulation of one of the most enduring tropes in the writing of scientific lives in India. In the subsequent sections I foreground the active collaborative roles played out by the scientists, Bose and Prafulla Chandra Ray in elaborating their self-images as rishis and *acharyas*, teachers, in terms of the Tagorean formulation, through two distinctive pedagogic projects that were meant to institutionalize and define the Swadeshi dispositions of Indian science.

In their attempt to do so I implicitly argue both Bose and Ray wittingly or unwittingly chose to ignore or disregard the

Science and the Making of a New Nationalist Masculinity   181

participation of Indian women in scientific pursuits. Although institutional structures and opportunities for the participation of women in science during this period remained deeply restricted both Bose and Ray were aware that women like Lady Abala Bose and Sister Nivedita were enthusiastic votaries of women's education and the pursuit of careers in western science and medicine for women.[12] Sister Nivedita was deeply supportive of Jagadish Bose's work and contributed greatly towards the publication of many of his works. Lady Abala Bose sought to actively pursue studies in medicine and remained consistently involved in Jagadish Bose's scientific work. Apart from them there were other women too who played important roles in the promotion of western science among Bengali women.[13]

It may be difficult to argue at this point that the creation of the one 'gendered world' of Swadeshi science was a deliberate act that chose to ignore the contributions of contemporary Indian/Bengali women in the larger Indian scientific endeavour, but the uncritical alignment of scientists and their patrons with the masculinist assumptions of Baconian science albeit in nationalist terms invariably closed the cultural spaces wherein Indian women could be imagined and acknowledged as equal participants.

In conclusion, I suggest a direction for developing the argument that the formulation of the image of the scientist as the bearer of a heroic masculinity took shape over time and through a dialogical engagement with gendered discourses of both Baconian science and cultural nationalist idioms of Swadeshi. The historical understanding of this critical alignment of science with the cultural and political discourses of nationalism offers new possibilities of interrogating received images of science and scientific practice in India, its gendering practices and the cultural and political legitimacy such practices have acquired over the years.

From the Improving Icon to the Swadeshi Hero

In much of what went by as writing on scientific lives in colonial Bengal, authors often remained loyal to genres developed in the West, particularly Victorian England.[14] What this meant was that such writing, much like its western counterpart, remained integrally related to a didactic project. Scientific lives, in other words, were written to elaborate notions of virtue or more comprehensively, 'character' which needed to be accepted and emulated by those to whom these were addressed. Central to this ideology was the belief that education was the key to 'improvement'—a catch-all term that acquired a particular resonance in the rapidly industrializing society of Victorian England (Hatcher 1996).

The Chamber brothers were the most popular precursors of Samuel Smiles (1812–1904) whose books *Self-Help*, *Character* and *Thrift* were the principal carriers of the creed of improvement in Victorian England (ibid.). The popular characters in this genre, though predominantly heroic, were not romantic heroes in that they seldom possessed supra-mundane qualities of imagination or ineffable genius. Instead, they succeeded because they possessed an ethical determination to pursue Truth.

The earliest writer of popular scientific biography, the Bengali scholar and social reformer Pandit Ishwarchandra Vidyasagar's first works *Betalpanchavimsati* and *Banglar Itihaas* were written under the patronage of Fort William College for the use of its European students (ibid.: 168). It was only with his writing of a model school book, based on the Chambers' 'Biography', translated as *Jivancharit* (1849) that Vidyasagar began to address a more explicitly Bengali readership. This revealed his larger pedagogic agenda. In his discussion of *Jivancharit*, Brian Hatcher observes that Vidyasagar's greatest innovation lay in the manner in which he affiliated the tenets of Baconian science and western bourgeois morality with an indigenous moral discourse of *nitishastra*. This was evident particularly in his ability to merge the virtue of industry with *yatna* (self-command) with *indriyajaya* (restraint of the senses), *ekagrata* (devotion) and so on (ibid.). It was through Vidyasagar's writings that the idea of science as a special kind of learning that had to be pursued with industry and devotion together with restraint and self-command came to be popularized.

In this rule of life enunciated by the improving school book scientists emerged as exemplars. The *Jivancharit* was quick to recognize this and included among its men worthy of emulation, scientists such as Copernicus, Galileo and Newton. In Vidyasagar's translated version the images of these men acquired new life. The biographical essay on Sir Isaac Newton was representative of the tone with which scientists and their practice was and would be addressed by Vidyasagar. Newton for example was lauded not only for his hard work (*satisay parisram*) but for being exceedingly diligent (*yatnavan*) in his researches on light and optics. For Vidyasagar, it was the combination of diligence and effort that led him to his discoveries and helped him receive the earthly rewards of fame and success (ibid.).

In his discussion of Vidyasagar's translated version of the Chamber's classic, Brian Hatcher thus observes that his greatest innovation lay in the manner in which he could affiliate the tenets of bourgeois morality, with the moral discourse of *nitishastra* in earlier Bengali school books. This was evident particularly in his ability to merge the virtue of industry with yatna, self-command with *indriyajaya* (restraint of the senses) and *ekagrata* (devotion) and so on. The pursuit of science in Vidyasagar's rendition of scientific lives was the pursuit of 'character' itself. His inclusion of the scientist in the discourse of Improvement tempered by the tenets of *nitishastra* had wittingly or unwittingly upheld the possibility of imagining the scientist as an emerging icon of Bengali masculinity (ibid.).

The rapid publication of successive editions of *Jivancharit* ensured for it a paradigmatic status in the genre of scientific biography. The notion that lives of scientists lent themselves to the elaboration of didactic texts extolling 'character' had become an orthodoxy that was seldom questioned. Vidyasagar's legacy was enriched and elaborated through the cultural and political constructions of Swadeshi science at the turn of the century. The incorporation of the scientist into the politic-ethical discourse of Swadeshi was facilitated by the poet Rabindranath Tagore's creative intervention as a friend and patron in the scientific career of Jagadish Chandra Bose (1858–1937). It was this cultural-nationalist framing of science that marked its alignment with emerging discourses of gender and class in Bengal.

Though frequently mentioned in accounts of the lives of both Tagore and Bose, the nature and complexity of their friendship has seldom been discussed in terms of its implications on the popular imagination of the scientist in colonial Bengal.[15] This needs emphasis because unlike the nature of science patronage which took shape in the context of the early middle-class enthusiasm for science in the 1830s and 1840s, this late nineteenth-century friendship generated an idiom of patronage, which had radical implications for the political image of science in the context of the nationalist movement. The beginnings of Indian participation in advanced scientific research transformed patrons from mere passive consumers of scientific knowledge into 'virtual' participants in the knowledge-making exercise.[16]

For Tagore, the friendship with Bose involved much more than the pleasures of affinity with another creative mind. It involved the steady and conscious accumulation of information on the experience of doing science, under conditions of colonial rule. Bose's letters to Tagore underlined the significance of scientific achievement in the context of material constraint, racial prejudice and intellectual isolation. For Tagore, patronage in these circumstances involved much more than material help; it involved an appreciation of the political and in a more profound sense the moral purport of the scientific enterprise.[17] His preoccupation with the political import of Bose's science found eloquent expression in verse, prose pieces and in innumerable review articles. Inspired by a nationalist imagination that had begun to see in the practice of science a route to cultural self-expression, these compositions lent a completely different inflexion to the rhetoric of science patronage at the turn of the century. In Tagore's literary renditions the pursuit of science transformed itself from a mundane scholastic enterprise into a supra-mundane intellectual quest, replete with romantic and moral affect.

His first composition on Bose, presented before a small gathering of family and friends in December 1897, was testimony to this new imagination. The verse with its opening line, '*Bigyan Lakshmir Priyo Paschim Mandire*' was first published in the Bengali periodical *Pradip* and later in one of Tagore's own collections *Kalpana*. A rapturous ode to a friend, this poetic response to Bose's success set the tone for Tagore's subsequent compositions. The lines were simple but profoundly evocative of sentiments that in a deep sense were shared by many of Tagore's contemporaries.

Dear Friend,
You ventured into the Goddess of Science's beloved western shrine,
And returned in triumph to garland the lowly head of your distressed motherland.
. . . .
Today she sends you her blessings through the voice of an unknown bard,
And hopes that its faint reverberations reaches and touches your heart.[18]

Apart from the fact that Tagore refers to science as a Goddess, the larger appeal of this poem perhaps lay in his imaginative combination of this with the iconography of *Bharatmata* (Mother India) to underscore the political import of Bose's success. The imagination of science as a goddess and a 'jealous goddess' at that, interestingly harks to Francis Bacon himself who made the connection in a work that sought to argue that the pursuit of science demanded ceaseless devotion to it.[19] It was an imagination that would be subsequently invoked by both Bose and Ray in different contexts.

As far as the imagination of the scientist as the 'nationalist son' was concerned, Tagore was evidently drawing upon the nineteenth-century image of Bharatmata which found its most sophisticated articulation in Bankim Chandra Chatterjee's *Anandamath* in 1882. There the image of the motherland as a widow in distress was elaborated through the role of the nationalist son as agent of deliverance. Bankim articulated in a remarkably persuasive manner what he perceived to be the duty of every nationalist son. This specific vision was to have a very special appeal for the later participants of the Swadeshi movement.[20] Tanika Sarkar has observed that 'for Bengalis accustomed to worship a wide variety of female cults, the emotional resonance connected with an enslaved mother tended to be particularly powerful'. The association of the scientist with the image of the nationalist son was also significant because of the perceived association of science with a exclusive masculine ethos.[21]

This was reinforced once again in a verse composition titled *Jagadish Chandra Bose* published a few months later. This was an extravagant celebration of the scientist as new India's first rishi. It was marked by its heightened tone, profusion of epithets and skilful deployment of the contrasting moods of rapture and reflection. Significant was the subtle transformation of Tagore's subject himself-from the young rishi contemplating truth in self-imposed isolation to an embodiment of the nation itself as the primal Guru or preceptor. Tagore's imagination it seemed had come full circle.

> *Where did you find your peace amid this din?*
> *Where did you find your silent nook*
> *Where you stood alone, with none other than the Great One himself?*
> ....
> *'Let all sophists and disputationists gather round you in submission,*
> *Let it redeem the seat of the Guru.'*[22]

The iconographic referents of the Bharatmata were jettisoned in favour of a more masculine imagination of the nation as the guru, radiating power in the form of wisdom. Although Tagore may not have been self-consciously aware of the sudden conceptual shift he was making in effecting a move from a feminine to a masculine imagery of the nation, it was clear that he saw no contradiction between the two. The projection of the nation–through the feminine mother icon was a way of introducing an affective tie to the nation while the projection of the nation through the masculine guru icon was a self-conscious display of an aspiration for power. For Tagore, Bose's commitment to his science was expressive of both the emotional and intellectual investment that was necessary to retrieve and reclaim the political and ethical foundations of the nation

Commenting on the singularity of Tagore's nationalist imagination, Dipesh Chakrabarty (2000: 167–68) argues that what lay at the heart of it was a distinctive *poetic vision*. In terms of his appreciation of Bose's science, the deployment of Tagore's poetic vision involved the act of looking at it in a manner that allowed him to see beyond the 'objective and historical' realities of Bose's context. The result was a certain transportation of a routine academic practice embedded in the 'everyday' of colonial rule into the realm of the transcendental. This, in turn, had the effect of rhetorically inscribing Bose's pursuit within a tradition of scholarship that was deemed to be pristinely Indian. In such a tradition the crown of the scholarly endeavour was believed to reside in the scholars' experience of the transcendental. By raising his scholarship to the level of this ancient

Indian ideal, according to Tagore, Bose had fulfilled not merely his aspirations and values but those of the nation.

While seemingly working within the discursive parameters of Bengali Hindu masculinity, I would argue that Tagore sought to transcend the sectarian bounds of Bankim's formulations by locating his ideal type not in any *puranic* or epic text but in the institutional spaces of colonial modernity. By imagining the scientist as rishi Tagore sought to both acknowledge and transcend the immediate presence of his male icon and formulate an ideal type that could cut across the bounds of religious communities. Given his cultural moorings in the Brahmo faith with its emphasis on a formless patriarchal godhead, Tagore's imagination of a heroic masculinity invariably pointed towards the imagined figures of the rishi and the acharya, both seemingly embodying a spiritual force of an ancient civilization. Sustained by the Orientalist dichotomies of East and West, Tagore's imagination of spiritual masculinity expressive of the combined strength of India's myriad faiths was meant to be a counterpoise to the aggressive, competitive, material masculinity of the West.

Tagore owed this belief, not merely to his empathy with the spirit of Bose's science, but to his intellectual affiliation to what is recognized as a pre-eminently nineteenth-century Bengali discourse on the pursuit of knowledge. Arguably, it was Bankim Chandra Chatterjee, literatteur and nationalist ideologue, who epitomized the struggle within the Bengali intelligentsia to resolve the conflicting pulls of the purportedly eastern and western ways to knowledge.

Partha Chatterjee (1997: 16–18), in his interpretation of Bankim Chandra Chatterjee's intervention, has argued that 'Bankim's resolution began with an explicit acceptance of the argument that the cause of India's poverty and subjection was the result of an absence in its culture of those attributes that had made the Europeans culturally equipped for power and for progress. In order to therefore qualify ourselves for progress and liberty it was necessary to transform our culture, to bring into being a new national religion suited to the modern world. The essence of this national religion would have to be a synthesis of the best achievements in the three aspects of knowledge: that of the world, the self, and God.' The implication of this would be an attempt to unite 'the European sciences' with 'Indian *dharma*'.[23]

The notion of *nishkama karma* as central to the making of the new Bengali self fed into the writings and speeches of Bankim's contemporary Swami Vivekananda who evolved a powerful discourse on Bengali masculinity in the later years of the nineteenth century by invoking the image of the *sanyasi*. In Vivekananda's reformulation, the sanyasi emerged as a political icon that embodied a balance between the active, passionate masculinity of the kshatriya with the sterner self-denying aspects of brahminical asceticism (Chowdhury 1998: 126). This icon of masculinity evidently evoked powerful sentiments in the minds of the nationalist intelligentsia, a social group increasingly marginalized by the operations of the colonial political economy and subjected to both overt and insidious forms of racial prejudice. Daily insulted by the realities of political subjection and yet powerless to hit back, this group found in the figure of the sanyasi the spiritual exemplar it needed to summon up 'from the depths of its soul the will and courage to deliver the ultimate sacrifice that would bring back the honour of the nation' (Chatterjee 1997: 18). Though seldom acknowledged, the political appeal of Bose's science was first noticed, by Swami Vivekananda at the 1900 Paris Universal Exhibition to which he too had been invited. Tagore was evidently inspired by the image of the 'courageous hero' which the swami saw in the young Bose and sought to endorse and sustain it, in the course of his engagement with his science. He recognized it as a nationalist endeavour because it embodied in its pursuit, the spirit of asceticism which both Bankim and Vivekananda had idealized. For Tagore, in its assertion of a distinctive masculine temper, Bose's science was an answer to racial stereotypes of the incompetent and 'effeminate' Bengali. His repeated invocation of the image of the rishi and the guru in the several verse-eulogies of Bose, were invocations of a new manliness defined not merely by the spiritual capacities of vision and an intuitive awareness of the transcendental but by a heroic capacity to bear hardships, undertake penance and make sacrifices. It was rhetoric of this kind that bestowed on Bose and the figure of the scientist with the virtues of what was described as 'practical asceticism'.[24]

The ultimate significance of Tagore's intervention lay in its marking the constitutive moment in the incorporation of modern science into an emerging discourse of cultural nationalism, which would in the context of the Swadeshi movement at the turn of the century, come to be reinforced through new modes of politics and new styles of creative expression.

The Scientist as 'Rishi'

Although Tagore could have been regarded as the author of the new role of the scientist as a masculine nationalist icon bearing the spiritual force of a purportedly Indian civilizational ethos, it was left to the subjects of his imagination to reinforce their self-images as rishis and acharyas in more compelling ways. While Bose took the initiative to 'sacralize' the space of his scientific research by establishing the *Basu Bigyan Mandir* (Bose Temple of Science) in 1917, his close friend and contemporary Acharya Prafulla Chandra Ray elaborated the inherent asceticism of the scientists' life in a popular rendition of his life in 1931. Although seldom seen in conjunction as critical inputs in the production of the hagiographic portrayals of Indian science, the establishment of the *Basu Bigyan Mandir* and the publication of the autobiography of Prafulla Chandra Ray served to reinforce the extra-scientific role of Indian scientists as seers and moral preceptors.

For Bose the role of the rishi was one he played out with unusual flair. His work on plant physiology may have discredited him in the professional world of western metropolitan science, but the response he received at home was enough to convince him of its moral and intellectual significance. His fourth scientific deputation abroad during 1914–16 brought him new recognition as the scientist of the East, one who could combine in himself the role of the modern scientist and the ancient seer (Dasgupta 1999). Bose chose to elaborate his vision and personality as a quintessentially 'Indian' scientist through the research Institute, the Bigyan Mandir he would establish in 1917.

Commenting on Bose's contributions to the elaboration of a nationalist scientific aesthetic, Ashis Nandy (1995: 80–81) writes that Bose proposed to establish a research institute that would be known as a Temple of Science. As emblems of the Temple, Bose selected sculptured representations of the Sun God (a symbolic pointer to the identification between *bhakti* or devotion and knowledge), *a vajra* or thunderbolt (the weapon of the Gods and the traditional symbol for legitimate fury) and the *ardha amalaka*, the Buddhist symbol of total renunciation. The Institute's interiors too were laden with symbolic effect. The lecture hall had a most remarkable piece of allegorical painting by one of Abanindranath's most famous pupils, Nandalal Bose. It depicted, as Jagadish Bose himself suggested, 'the idol of knowledge floating down the Ganges with the eternal woman beside him–a representation of *Shakti* inspiring *Purusha*' (ibid.). Although Nandy refrains from explaining the significance of this iconography it pushes one to ask if this Hindu iconography of Shakti inspiring Purusha was reflective of Bose's attempts to develop a more inclusive ideal of scientific practice or if it merely elaborated the idea of the Goddess of Science (Shakti) inspiring a male devotee (Purusha).

In a speech delivered in Calcutta shortly before the final touches were being put to the plans for the *Bigyan Mandir*, Bose proclaimed that there would soon rise a 'temple' where the teacher cut off from worldly distractions would go on with his ceaseless pursuit of truth and, on dying, hand over his work to his disciples. 'Nothing would seem laborious in his inquiry, never was he to lose sight of his quest. For his was the *sanyasin* spirit' (Ghosh 1931: 4). For Bose this was the quintessentially 'Eastern spirit' in which knowledge was pursued neither as a social attribute nor as a route to material advantage but as a way that led to a deeper understanding of the self and of God. Knowledge understood in these terms had been cultivated in the spirit of a sanyasi or renouncer.

The shorthand, which Bose used to describe it, was the 'Eastern spirit', and one of the earliest occasions in which he spoke about it was in a speech he delivered before the Bengal Literary Conference in Mymensingh in 1911.[25] Here the 'Eastern spirit' was invoked in order to dissent against the norms of western scientific practice, its fetish for specialization and its narrow utilitarian concerns. Tagore composed a memorable song for the occasion, the inspiring lyrics of which set the tone and ambience for the inauguration of the Bigyan Mandir. Indeed, the association of the research institute with the sacred space of a temple, the infusion of research itself with the values of righteousness, virtue and heroism, and the patriotic vision of national glory through scholarly attainment were to be the central themes of the founder's dedication too. Bose's speech entitled the 'Voice of Life', published as a monograph, was one of the most important documents on the philosophy and ethics he sought to bring into the practice of science in India. It recorded his sense of unease with the norms of western academic practice, his chance discovery of the underlying unity pervading the natural world, the story of his tortuous struggle to establish the claims of this fundamental truth and his profound sense of joy at being able to bequeath to the nation a possible path to greatness and honour.[26] Writing at a time when the Great War generated a deeper sense of pessimism about the progressive, emancipatory potential of science, Bose sought to invest the claims of the Eastern spirit with a new language of legitimacy.

For Bose, the spirit of renunciation was as integral to life as it was to any kind of scholastic endeavour. One of the primary objectives of the Bigyan Mandir was the creation of those conditions in which at least a chosen few could revive the struggle to attain that ideal. 'I call on those very few' he proclaimed, 'realizing some inner call, will devote their whole life with strengthened character and determined purpose to take part in that infinite struggle to win knowledge for its own sake and see truth face to face.' He concluded 'With this widened outlook it would be possible not merely to maintain the highest traditions of the past but also serve the world in nobler ways feeling the common surging of life, the common love for the good, the true and the beautiful' (ibid.: 588).

Whether or not Bose actually adhered to this idealized self in his professional or private life remains open to debate. But his much-publicized struggles were indicative of his predicament as a scholar, while his choice of the icon of the 'renouncer' or sanyasi as moral exemplar was proof of his commitment to this ideal. It reflected his concerns about both the requirements of scientific research in the country and the larger and perhaps more fundamental preparation for the struggle against colonial rule. Asceticism as an ideal was as essential to scientific scholarship as it was to the cultivation of a larger spirit of resistance. The Bigyan Mandir was meant to convey this deeper nationalist impulse. It signified Bose's own perceptions of his style of scientific practice as well as his evolving affiliation with the aspirations of Bengali middle-class nationalism. The accent on struggle and renunciation as the ideal and ethic of scientific practice was his way of endorsing and transcending the political discourse of asceticism as it took shape in the writings of Bankim Chatterjee, Swami Vivekananda, Sri Aurobindo and the ideologues of revolutionary terrorism in the Swadeshi years.[27] In all these writings the asceticism was represented as an attribute, which generated the will to resist. By associating the cultivation of science with the cultivation of the virtue of asceticism the Bigyan Mandir and its ideology of scholarship legitimized the tenor of the rhetoric of science patronage with the reassertion of the belief that the identity of nationalist science was defined by the nature of the moral comportment that the scientist brought to bear upon his practice.

Towards the close of the second decade of the twentieth century, the rhetoric around the 'Indian-ness' of Indian science had achieved a degree of coherence. Informed as it was by prevailing trends of aesthetic discourse in Bengal and linked closely to the power and influence of Orientalist knowledge, this rhetoric of science and nationalism helped nurture the self-image of Indian scientists and generate a sense of community marked by a distinctiveness in its ethical outlook. While Bose and Tagore drew upon the ideals of Bankim and the early Swadeshi ideologues to foreground the ethical imperative of renunciation, nationalist scientists in the 1930s and 1940s drew out the social implications of science practised on its basis. In the context of the ideological effects produced by the Gandhian intervention in nationalist politics it was left to Bose's contemporary Prafulla Chandra Ray to effect the subtle displacement of the 'Orientalist ideal' of Indian science with that of a newly emerging social ideal premised on a similar ascetic sensibility.

The Practical Ascetic

Scientist, nationalist, Bengali *bhadralok*, the western-educated, generally upper caste and middle or upper class, intellectual, entrepreneur, public figure, sometime Gandhian, almost-politician, nearly-socialist, Prafulla Chandra Ray was a major influence on the scientific fraternity in India, giving them a legitimate voice as Indian scientists and the confidence to practice in a new and less unequal environment. He was deeply involved in the debates on Indian nationalism from the late nineteenth century to independence.[28] Little wonder that when it was published in 1932, Ray's autobiography, *The Life and Experiences of a Bengali Chemist*, presented itself as something more than an autobiography (Ray 1996). In the author's words, it was meant to be a 'lesson' for his countrymen. He considered his life to be exemplary and worthy of emulation, for three compelling reasons: One, because he was convinced that it was a life which had served science and country with equal devotion; Two, because it was a life that to him, exemplified the discipline and spirit which Bengali youth needed to cultivate to retrieve its dignity. And finally, because it was a life that had demonstrated by example that the ethical imperative of Swadeshi required both cultivation and transcendence of individualist aspirations in the name of a greater collective: the nation. By framing and validating his 'lesson' in these terms, Ray to all appearances, sought to do two things at once—to resolve his own anxieties relating to the nature of his affiliation to the bourgeois/individualistic ethos of modern science and to validate the claims of the scientist as moral exemplar.

In *Life and Experiences* (in particular the Bengali version *Atma Charit*), Ray's invocation of God and the Self sought to transcend that individualism in the name of community and nation:

> It is unfortunate that in one's own memoir the frequent use of the first person singular cannot be dispensed with; one naturally lays oneself open to the reproach of egotism. An awful burden of responsibility has borne me down, whenever I have had to use it. Whatever field I have ploughed I have ploughed as an humble instrument in the hands of Divine Providence; my failures are my own; to err is human. But my successes if any are to be attributed to the guidance of the All-knowing, who chose me to be His humble servant. After all, a Divinity shapes our ends . . . I thought however that a plain unvarnished narrative of my uneventful career which has run its noiseless tenure, might convey some lesson to my countrymen, specially of the younger generation (ibid.: 541).

Ray's moral framing of the Self and God was in many ways premised upon the inclusion of Society. The moral exemplariness of himself as subject of his autobiography, in other words, had to be located not at the point of an intense individualistic relationship with God but in the deep engagement with Society as a result of it. Ray's self-projection of the scientist, thus, differed from that of Bose. The spiritual perfection that Bose sought to attain through his science was an individualistic quest that needed the private space of a secluded environment. Ray's moral perfection, however, demanded the ascetic attributes of Bose's sanyasi-scientist but only in reference to a life devoted to a puritanical ethic of work.

Of interest to his readers may have been the number of ways in which Ray engaged with the term 'work' itself. This revealed the complex trajectory of Ray's engagement with what was purportedly one of the central tenets of the 'improving' ethic. It was, in a sense, critical to the lesson he sought to impart to his countrymen, For if the nation had to be recovered through work, or if Bengali men had to retrieve their dignity through work, they had to address themselves to the political and moral implications of work. Readers were thus presented with not only what amounted to a roster of Ray's life works, that is, deeds accomplished in his lifetime, but to the various moral connotations of the term, as it figured in his general discussions on attitudes to physical work or labour, on professional and constructive work, on work as leisure, as it related to time, and finally as social service. Evident in this intense preoccupation with the notion of work, was perhaps Ray's attempts to engage the bourgeois-individualist assumptions of 'work' in the received discourse of 'improvement', and to redefine it in terms of an alternative notion of work that had taken shape in the nationalist discourse of Swadeshi. From all his discussions, it becomes clear that work for Ray, ultimately meant a sustained commitment to what in the Swadeshi lexicon stood for 'constructive work' or work informed by the self-transcending ethic of 'service'.[29]

In the elaboration of this self-transcending ethic of service, Ray owed much to the influence of Mahatma Gandhi. Though often a source of controversy, the Gandhian connection for Ray was significant for the impact it had on his self-image outside academia. The two had met several years before the 1920s, and Ray had apparently come to admire him almost immediately. In the chapter titled 'Life outside the Laboratory', he chose to document the transformation in his own persona, as it came to be subtly moulded in the Gandhian cast. Critical in this transformation was his wholehearted endorsement of the message of khadi. Ray's advocacy of the Gandhian programme, however, earned him the reproach of friends and admirers. Deriding his zeal for the charkha, Brajendranath Seal invested him with the dubious distinction of being a *charkarishi* (an amalgamation of *charka* and *rishi*).[30] Although Ray took such criticism in his stride, he strove hard to convince his readers that what underlay the seeming contradictions of his life was his unwavering moral conviction. He began his conclusion with an apology:

> Some might at the very outset offer the advice that a shoe-maker had better confine himself to his task.; that a chemist had no business to go beyond the range of his laboratory. Fortunately or unfortunately as the case maybe The subject of this memoir has been something more than a mere chemist. I am what I am and cannot help being made up of perhaps incongruous elements .... Whether the materials presented here are of an incongruous nature strung together or whether they bear some sort of relation to the career of the Bengali chemist it will be for the reader to judge. (Ray, vol. 2, 1996: 539–40)

In the very admission of this imperfection, Ray sought recognition of the perfection of his moral position. He had held his life to the scrutiny of his countrymen and it was for them to judge if he had lived a life worthy of emulation. By elaborating the possibility of moral rectitude and simultaneously expressing diffidence in attaining it, Ray was introducing what he described as his ordinary life into a life of virtue.

Like many other nationalist preoccupations with science in the twentieth century, Ray's autobiography too remained focused on the moral disposition of the scientist rather than on the knowledge he produced. To his readers the nationalist assumptions of *The Life and Experiences* was evident in that in the final analysis it was as much a life-story of a Bengali scientist as it was the story of Bengali science. By the mid-1920s he had discovered an alternative role: that of the Gandhian. In adopting that role and fashioning himself as both an 'improving icon' and a Gandhian moralist, Ray seemed to have succeeded in defining both his own image and that of Indian science at large.

In his writings, Ray wove in Victorian attributes of masculinity but presented its central message as a distinctively nationalist one. The singular import of his autobiography ultimately lay in its feeding into and lending credence to an idealised image of the Indian scientist as the bearer of a civic nationalist commitment that was powered in by a heroic masculinity expressive of a Calvinist devotion to work, a Smilesian motivation to achieve and the Swadeshi ideal to serve the nation with an ascetic disposition.

The upshot of Ray's pedagogic project much like Bose's had the effect of reinforcing an image of the scientist already in circulation in Bengal's literary public sphere. Bengal's men of science, for instance, acquired new visibility in the *Vanga Gaurav Granthamala* series dedicated to popular renditions of the lives of the good and great of Bengal. *Bigyane Bangali*, a part of this series, was a collective biography of Bengali scientists and one that claimed to be the 'first illustrated account of the lives of Bengal's heroic men of science' (Ghosh 1931: 1). Together with similar life sketches of Bengal's statesman, warriors, sages and scholars, it was meant to complete the pantheon of Bengal's masculine heroes. Its author Anil Chandra Ghosh (ibid.) believed that the book would imbue timid minds with the spirit of adventure and inspire them to master the methods of the sciences for the glory of the nation and the honour of its people.

Bigyane Bangali in some ways was indicative of the ways in which the discourse of practical asceticism rendered the cultural fit between science and masculinity. For all its apparent simplicity, *Bigyane Bangali* could be effectively situated in the multiple contexts in which science operated in colonial society and the numerous ways in which its authority was culturally redefined and re-instituted. Its use of an eclectic range of literature revealed the terms in which it affiliated science with a distinctively masculine enterprise inspired by and embedded in the cultural and political project of Swadeshi.

Ghosh's moral logic worked to impose a certain pattern on the text. All his protagonists were seen to undergo similar experiences of trial, tribulation and ultimate triumph in their lives and all of them were known to have displayed similar virtues of dedication, sacrifice and a sense of duty to attain their objectives. All of them were believed to have succeeded in their vocations because of their 'ascetic' dispositions. They were ultimately seekers of truth who shied away from material rewards and cultivated habits of renunciation to fulfil their aspirations. The upshot of this exercise was, in no uncertain terms, an idealization of the Indian scientific enterprise in terms of its intrinsic ethical disposition.

Bigyane Bangali, for all its claims to originality however, was not an isolated text. Indeed, it was a redacted version of *Indian Scientists,* a collection of scientific lives in English published by G. A. Natesan and Company in Madras in 1929 as part of a series on the lives of eminent Indians.[31] The inclusion of scientists in a whole range of nationalist icons was indicative of the kind of image the profession had come to acquire in little more than a decade of its emergence at the turn of the century. *Bigyane Bangali* included life-sketches (in order) of Mahendra Lal Sircar, J. C. Bose, P. C. Ray, C.V. Raman, Srinivasa Ramanujan and Professor Ramchandra. The select list was carefully prepared in order to make the volume truly representative of the Indian endeavour. This is where the volume announced a self-conscious departure from its predecessor. In its attempt to privilege the Bengali scientific endeavour, it deliberately included the lives of scientists from the south. In all other aspects of style, content, format and the much publicised didactic intent, the two books were remarkably similar.

It was perhaps fortuitous that *Bigyane Bangali* was brought out a few months before *Life and Experiences.* When read together, both revealed the extent to which the cultural discourse on Indian science that took shape in the context of Swadeshi was sustained and reinforced in the domain of literary representation. P.C. Ray's 'lesson' for his countrymen with its accent on service and sacrifice fed into the effusive celebrations of the moral predicament of Indian science in texts like *Bigyane Bangali* to reinforce the cultural nationalist argument, that the identity of Indian science lay in the distinctively

Indian ethic of its practice. It was left to an emerging nationalist leadership of the mid-1930s to draw upon this image and redefine it in terms of its own specific political agenda.

Conclusion

Notwithstanding the years that separate nationalist science in colonial Bengal with scientific practice in contemporary times, there remains an unceasing connection between the conventions of writing lives in science in the early twentieth century and those written in more recent times. The connection lies both at the level of rhetoric as well as intent. Science and the virtues of heroic masculinity that purportedly inform it, are part of what has been described as a 'sacred and natural' history of nationalism.[32]

By focusing on the active roles played by scientists and their patrons in elaborating, propagating and legitimising this history, my attempt is to raise questions about the political implications of this exercise and the exclusionary implications it has for women's visibility in science in the larger public domain. The constant and deliberate inclusion of representations of Indian science within what is largely a gendered history of nationalism resists attempts to raise questions regarding the historical and political underpinnings of such an endeavour. Questioning the legacies of the early linkages between science and the cultural politics of colonial middle-class nationalism offers a way of reinstating the political questions that have remained suspended in academic debates at the moment of independence and thereafter. Although the linkages between science and the gendered discourses of nationalism were made in exceptional circumstances, it may be worth asking why several years after those exceptional circumstances and after the routine participation of women in the scientific profession and in the nation-making project at large, the self-representations of Indian science remain unproblematized.

By attempting to decode the cultural and political underpinnings of the self-representations of Indian science this essay has sought to forge alliances with the larger ongoing debates on gender and science in the Indian context. It has sought to argue that the story of

Science and the Making of a New Nationalist Masculinity   199

Indian science needs to be situated in histories that have explored the many complexities of gender relations in colonial India and the specific framing of nationalist masculinities.[33] By doing so, it has tried to contribute towards the larger exercise of interrogating received images of the scientific endeavour and of locating the historically specific contexts within which questions relating to the equitable participation, access and visibility of women in science were rendered irrelevant.

Notes

1 Pioneers in feminist historical readings of early modern science are Carolyn Merchant (1980) and Evelyn Fox Keller (1985).
2 For an early discussion of colonial masculinity and nationalist responses to it see Mrinalini Sinha (1995) and Indira Chowdhury (1998) among others. Both authors examine the centrality of gender in the project of colonial rule and the formation of national identity. Chowdhury in particular explores the ways in which indigenous cultural forms were reworked for the articulation of a modern political identity.
3 For a fuller explication of the argument see Jan Golinski (2002: 125–45).
4 Other alternative readings of Bacon's misogynist ideology of science include, Sarah Hutton (1997) 'The Riddles of the Sphinx: Francis Bacon and the Emblems of Science'. In Lynette Hunter and Sarah Hutton, eds (2007).
5 See for instance, John Lourdusamy (2004).
6 Mahendralal Sircar (1833–1904), medical practitioner, championed the promotion of scientific research among Indian nationals; established the Indian Association for the Cultivation of Science in 1876.
7 In the mid-nineteenth century what was meant by the patronage of science in Bengal involved little in terms of patronage of scientific *practice*. Patrons of science then had their interests focused on creating a sense of wonder and curiosity around the new knowledge together with an appreciation of the usefulness of such knowledge to their lives. Inspired by the 'rationality' and 'improving vision' of western science, mid-nineteenth century science–enthusiasts, many

of whom belonged to the *Brahmo Samaj*, evolved elaborate moral defences in favour of scientific learning, particularly, in terms of the bearing it had on their project of social reform. In the absence of a knowledge of the experiential realities of practicing science under colonial conditions or of the politics that determined the *content* of the knowledge available to them, patrons of science in the mid-nineteenth century retained faith in the gentle processes of diffusion of western science in the Indian context. There was only if at all, an intuitive awareness of the complexities of the political struggle, necessary to engage with western science on their terms. By the late nineteenth century, however, what was evident was a remarkable transformation in the terms of patronage as a result of the beginnings of an Indian participation in scientific research. While this has seldom drawn critical attention, the event of the Indian participation in scientific research, generated an entirely different engagement with science. It is here, that the specific idiom of patronage that came to be elaborated through the friendship between the poet Rabindranath Tagore and the scientist Jagadish Chandra Bose came to represent a departure.

8 Gyan Prakash (2000) brought together in a coherent manner his position on the cultural terms in which modern science was received in India. In his earliest statement on the subject (1996), he drew attention to Prafulla Chandra Ray's *History of Hindu Chemistry* and Jagadish Chandra Bose's theory on plant response, as evidence of the way in which a Bengali intelligentsia sought to engage with modern science in terms of its own cultural and political aspirations. This early statement was later developed with a more comprehensive survey of the nationalist engagement with science in the later book.

9 See Partha Chatterjee (1993: 6–7).

10 Later, however, Chatterjee did go on to acknowledge the presence of science in the nationalist cultural project. In a lecture titled 'Our Modernity' in 1994, which was published in a collection of essays (Chatterjee 1997: 207–08) he said, 'We have of course seen many attempts ... in the fields of literature and arts to construct a modernity that is different. Indeed we might say that this is precisely the cultural project of nationalism: to produce a distinctly national modernity.... My argument was that these efforts have not been restricted to the supposedly cultural domains of

religion, literature and the arts. The attempt to find a different modernity has been carried out even in the presumably universal field of science. We should remember that a scientist of the standing of Prafulla Chandra Ray, a Fellow of the Royal Society, thought it worth his while to write a *History of Hindu Chemistry*, while Jagadish Chandra Bose, also an FRS, believed that the researches he carried out in the latter part of his career were derived from insights he had obtained from Indian philosophy.'

11 In the historical literature on the significance of the 'moment' of Swadeshi in the cultural politics of Bengali middle-class nationalism, it has been argued that what invested the idea Swadeshi with both coherence and appeal was the convergence of its civic discourse with its conception of an oppositional cultural identity. Both were seen to cohere on the issue of the ethical imperative of 'asceticism'. The notion of a 'this-worldly asceticism' which began its career in the writings of Bankim was critical to the elaboration of both the economic and moral order of Swadeshi. In the writings of most Swadeshi ideologues it was variously argued that the ideal of asceticism was valorized for its import in defining not merely the non-acquisitive ethos of Swadeshi economics but for its capacity to generate the "spiritual force" necessary to reclaim for the Bengali middle-classes both their manhood and their moral being. See Sartori (2003) and (Chowdhury (1998).

12 Lady Abala Bose (1865–1911), wife of Jagadish Chandra Bose, a social reformer and advocate of women's education and women's franchise. Denied admission to Calcutta Medical College, she proceeded to Madras for medical education but was unable to complete the course. One of the early Brahmo women along with Kadambini Bose who remained enthusiastic votaries of the inclusion of science, philosophy and mathematics in women's education and sought careers in medicine for them. For an overview of women's education in colonial Bengal from the 1850s to the 1920s see among others, Rachana Chakraborty (2009). For a review of the debates on the contributions of Sister Nivedita (1867–1911) to Bose's scientific writing, see Siladitya Jana (2013).

13 I refer here in particular to the writings of Begum Rokeya Sakhawat Hossein: writer, social reformer, advocate of women's education. Her feminist utopia *Sultana's Dream* written in 1905 describes the deft use of science and technology by women in their bid to break out of their patriarchal social confines and stereotypes.

14 For a discussion of this genre and its impact on genres of didactic literature in nineteenth-century Bengal, see Brian Hatcher (1996: 117–27). For a detailed comment on Vidyasagar's pedagogical project, see Sumit Sarkar (1997): 256–57.
15 Patrick Geddes (1920); Prasanta Kumar Pal (1990); and Subrata Dasgupta (1999). All three, for instance, talk about the friendship but never on its larger cultural consequences. The term 'patronage' used here, derives from the understanding of patron as 'a person, group or organization which gives support, encouragement, or financial aid: an upholder'.
16 For a discussion of the new modes of scientific knowledge making and the audiences for it see Jan Golinski (2000).
17 Here it may be noted that Tagore was astute enough to realize that material help for Bose's work could be garnered only if its ethical purport could be underscored. This was particularly true in a context where the complexities of Bose's research were little understood.
18 This is a shortened version of the original. Translation mine.
19 A point made by Sarah Hutton (1997).
20 See discussion in Chowdhury (1998: 95–118).
21 For a detailed discussion on the subject see Ludmilla Jordanova (1989: 5). In this she suggests that 'the very terms in which knowledge, nature and science are understood are suffused with gender, partly through the tendency of ancient origin to personify them and partly because as a result we think of the processes by which knowledge is acquired as deeply sexual'.
22 This is a much shortened version of the original titled, 'Jagadish Chandra Bose' and published in *Bangadarsan* (mid-June to mid-August), Ashadha-Sravan (mid-June to mid-August) 1897 and reprinted in P.B. Sen, vol.6 (1974: 99–100). Translation mine.
23 See discussions of Bankim Chandra Chatterjee's elaboration of the foundations of 'national culture' in Partha Chatterjee (1986: 54–84).
24 For Vivekananda's visit to Paris and his comments on Bose, see Banhatti (1995: 192–93).
25 For a discussion of the controversial reception of Bose's science in the West, see among others (Dasgupta 1999: 55–59).
26 This address was one Bose's most eloquent statements on the philosophy and ethics of his science. The extract of the address used in

Science and the Making of a New Nationalist Masculinity 203

this chapter is taken from an English version published in the *Modern Review* (6 December 1917: 585–95).
27 For a detailed discussion see Indira Chowdhury (1998: 120–49).
28 Acharya Prafulla Chandra Ray's nationalistic writings go back much earlier in his career. He wrote extensively on various subjects in almost all contemporary journals. He began his literary career with his prize-winning essay 'India Before and after the Mutiny' written while still a student at Edinburgh. Back in Calcutta, his first major literary piece came out in 1901. Titled *Charak o Shushruter Samay* (The Times of Charaka and Sushruta) this piece was a kind of prequel to his major literary work that was published in the following year. The first volume of the *History of Hindu Chemistry* came out in 1902 and acquired international acclaim. A host of shorter pieces on the history of science in ancient India followed thereafter. The year 1910 marked the beginning of a different literary preoccupation with the publication of the tract entitled *'The Bengali Brain and Its Misuse'*. Ray's literary and journalistic skills from this moment on remained focused on a singular didactic project: the political and economic regeneration of the Bengali. His autobiography of 1932 was a logical sequel to his previous literary and political preoccupations. All references from his autobiography *Life and Experiences of a Bengali Chemist* used in this chapter are from reprints of vols 1 and 2 brought out by the Asiatic Society, Calcutta, 1996.
29 Memorable quotes include, 'When work is coupled with a keen sense of enjoyment, it does not tell upon one's health', *Life and Experiences*, vol.1: 106, 'Work, i.e., congenial work is pleasure': 213, 'Activity is the only road to Knowledge': 544, 'Work itself was my delight and as I experienced almost a romantic sensation. I did not break down or let my interest flag': 84 'Work in connection with social service has always been my hobby': 85 'Constructive work either in connection with my own science and its application to industry or in relation to the removal of economic distress has always engrossed my chief attention': 225.
30 Brajendranath Seal (1864–1938), Ray's illustrious contemporary-scholar, philosopher, member of the Brahmo Samaj and Vice-Chancellor of the University of Mysore. Letter from Brajendranath Seal to Ray, 12 September 1928, Mysore, P.C. Ray Private Papers, Nehru Memorial Museum and Library, Delhi.

31 G. A. Natesan (1929). The Natesan volume on *Indian Scientists* was significant because of the immense respect enjoyed by this publishing house in Madras. Its owner, G.A. Natesan, belonged to the liberal elite of Mylapore with close connections with the Indian National Congress. It was under Natesan's personal initiative that a series of popular texts on famous Indians were brought out since the 1920s. The links between *Bigyane Bangali* and *Indian Scientists* are important in this respect because both texts were seen to have been drawing upon the cultural nationalist discourse of Swadeshi in their attempts to draw out the nationalist credentials of Indian science.

32 For an elaboration of this idea, see among others Pandey (2006), chapter 5: 103–28.

33 There is a rich and growing body of literature on the multiple nuances of the nationalist masculinist paradigm that try to show how complex and plural paradigms of masculinity that existed in the subcontinent were often jettisoned in favour of hyper-masculinist models in the context of specific nationalist projects. Studies of how scientists in their own self-representations straddled the brahmanical and kshatriya models of masculinity or offered more nuanced models remain to be done.

References

Banhatti, G.S. 1995. *Life and Philosophy of Swami Vivekananda*. New Delhi: Atlantic Publishers.

Chakrabarty, Dipesh. 2000. *Provincializing Europe: Postcolonial Thought and Historical Difference*. Princeton, NJ: Princeton University Press.

Chakrabarty, Rachana. 2009. 'Women's Education and Empowerment in Colonial Bengal'. In *Responding to the West: Essays in Colonial Domination and Asian Agency*, edited by Hans Hagerdal. Amsterdam: Amsterdam University Press: 87–101.

Chatterjee, Partha. 1997. *The Present History of West Bengal: Essays in Political Criticism*. New Delhi: Oxford University Press.

———. 1993. *The Nation and its Fragments: Colonial and Post-Colonial Histories*. Princeton, NJ: Princeton University Press.

Chowdhury, Indira, 1998. *Frail Hero, Virile History: Gender and the Politics of Culture in Colonial Bengal.* New Delhi: Oxford University Press.

Cohn, Carol. 1993. 'War, Wimps and Women'. In *Talking Gender, Thinking War: Gendering War Talk,* edited by Miriam Cooke and Angela Woolacott. Princeton, NJ: Princeton University Press 225–46.

Dasgupta, Subrata. 1999. *Jagadish Chandra Bose: An Indian Response to Western Science.* New Delhi: Oxford University Press.

Geddes, Patrick. 1920. *The Life and Work of Sir Jagadish Chandra Bose.* London: Longmans, Green.

Ghosh, Anil Chandra. 1931. *Bigyane Bangali.* Dhaka: Dhaka Library Press.

Golinski Jan. 2002. 'The Care of the Self and the Masculine Birth of Science'. In *History of Science,* vol. 40: 125–145.

———. 2000. *Making Natural Knowledge: Constructivist Approaches to the History of Science.* Cambridge, Cambridge University Press: 1–12.

Harding Sandra.1991. *Whose Science, Whose Knowledge? Thinking from Women's Lives,* Ithaca, NY: Cornell University Press.

Hatcher, Brian. 1996. *Idioms of Improvement: Vidyasagar and the Cultural Encounter in Bengal.* New Delhi: Oxford University Press.

Hunter, Lynette, and Sarah Hutton, eds. 1997. *Women, Science and Medicine 1500–1700: Mothers and Sisters of the Royal Society.* Stroud, Sutton Publishing.

Keller, E. Fox. 1985. *Reflections on Gender and Science.* New Haven, CT: Yale University Press.

Jana Siladitya, 2013. 'Sister Nivedita's influence on J. C. Bose', *Journal of the Association of Information Science and Technology.* Wiley Online Library, DOI 10.1002/asi.23221: 1–6.

Jordanova, Ludmilla. 1989. *Sexual Visions: Images of Gender in Science and Medicine between the Eighteenth and Twentieth Centuries.* Madison, WI: University of Wisconsin Press.

Lourdusamy, John. 2004. *Science and National Consciousness in Bengal: 1870–1930.* New Delhi, Orient Longman.

Merchant, Carolyn.1980. *Death of Nature: Women, Ecology and the Scientific Revolution.* San Francisco: Harper and Row.

Nandy, Ashis. 1995. Alternative Sciences: Creativity and Authenticity in Two Indian Scientists. New Delhi: Oxford University Press.

Natesan, G.A. 1929. Indian Scientists. Madras: G.A. Natesan and Company.

Pal, Prasant Kumar. 1990 *Rabijibani*, vol. iv. Calcutta: Ananda Publishers.

Pandey, Gyanendra. 2006. *Routine Violence: Nations, Fragments, Histories.* New Delhi: Permanent Black.

Gyan Prakash, 2000. *Another Reason: Science and the Imagination of Modern India.* New Delhi: Oxford University Press.

———. 1996. 'Science between the Lines'. In *Subaltern Studies, IX, Writings in South Asian History and Society,* edited by Shahid Amin and Dipesh Chakrabarty, New Delhi: Oxford University Press: 59–82.

Ray, Prafulla Chandra. 1996 (rpt). *Life and Experiences of a Bengali Chemist,* vols, 1 and 2. Calcutta: Asiatic Society.

Sarkar, Sumit.1997. Vidyasagar and *Brahminical* Society', in *Writing Social History.* New Delhi: Oxford University Press.

Sarkar, Tanika. 1987. 'Nationalist Iconography: Images of Women in Nineteenth-Century Bengali Literature', *Economic and Political Weekly* 22, 47 (21 Nov.): 2011–15.

Sartori, Andrew. 2003. 'The Categorical Logic of a Colonial Nationalism: *Swadeshi* Bengal, 1904–1908'. In *Comparative Studies of South Asia, Africa and the Middle East* 23, 1–2: 271–85.

Sen, Pulin Behari, ed. 1974. *Chithipatra,* vol. vi. Kolkata: Vishwa Bharati.

Sinha, Mrinalini. 1995. *Colonial Masculinity: The 'Manly Englishman' and the 'Effeminate Bengali' in the Late Nineteenth Century,* Manchester and New York: Manchester University Press.

Vivekananda, Swami. 1989. [1922]. 'Memoirs of European Travel'. In *The Collected Works of Swami Vivekananda*, vol. vii. Kolkata: Advaita Ashram: 379–80.

22

Fingerprints and Erasures: Mapping the Creative Process in Science

Gita Chadha

All versions of the scientific world-view take science to be a totalizing theory; it has been assumed that anything and everything worth understanding can be explained or interpreted within the assumptions of modern science. Yet there is another world hidden from the consciousness of science—the world of emotions, feelings, political values; of the individual and collective unconscious, of social and historical particularity explored by novels, drama, poetry, music and art—within which we all live most of our waking and dreaming hours under constant threat of its increasing infusion by scientific rationality. Part of the project of feminism is to reveal the relationship between the two worlds and how each shapes and forms the other (Harding 1986: 245).

The broad aim of this chapter is to further the critical approaches to modern science that have emerged in the last forty years from 'outside' science from the social sciences and social movements. These approaches have challenged the conventional representation of science as an asocial and ahistorical knowledge-making system. They seek to explain science as a socio-historical process and a carrier of cultural formations. They challenge the notion that science is an autonomous, universal and neutral knowledge-making system. These approaches seek to challenge the notion that science is inclusive of women, people of colour and other marginal groups, that science is simply determined by notions of merit.

Specifically, based on narratives of scientists, I present an analysis of the creative process of doing science across different disciplines. It maps the various stages of knowledge-production in what are called the 'fundamental' sciences. It also unpacks the notions of the 'genius' in science, and assumes that a critical understanding of these processes reveals the social and cultural aspects that are marginalized and erased in the mainstream discourses of science. The framework is informed by a triad of theoretical approaches: the new sociological studies of science, the feminist approaches to science and the postcolonial discourses about science. The three approaches, in a sense, represent my own locations as a social scientist, a woman and a postcolonial subject. The journey through these frameworks is, therefore, as much personal as academic.

The Theoretical Triad: From Sociological, Feminist and Postcolonial Locations

Originating in the Enlightenment thought of western modernity, the social sciences, particularly sociology, aimed at using reason—represented by modern western science—to understand the social world, making the social-science project almost derivative of the natural sciences. While methodological debates initiated by the interpretivist sociology of Max Weber at the time did question the adequacy of using natural science as a model for the social sciences, sociological positivism emerged as the dominant paradigm within sociology leading to the hegemony of natural science in general sociology. The sub-disciplines of the sociology of science and the sociology of knowledge that were to focus specifically on studying science, merely took the 'weak' approach of looking at the institutional aspects of science rather than at scientific knowledge as being socially determined. Canonically, sociology of knowledge is expected to explore socio-cultural contexts within which knowledge systems like philosophical, religious, political and aesthetic systems are produced. Science was excluded from this gaze. The central question that begs an answer is: do not all forms of knowledge need be subjected equally to a sociological scrutiny or do we exempt some? Whatever may be our own preferences as sociologists, does not our discipline require us to look at science with as much rigour as we would look at other knowledge-making systems for sociological insights? (Bloor 1976). The pioneers of sociology of knowledge like Karl Mannheim (1936) accepted the position that science as a system of knowledge is not influenced by the social, indeed, that the scientific method functions like a sieve to remove 'social errata' in its truth-making efforts thus according science a privileged and special status as a knowledge-making system. This position undoubtedly strengthened the hegemony of science within sociology. This position adopts what has been called the 'standard view of science' (Ziman 2000: 28), wherein the reality and the objectivity of the physical universe and the subject matter of science are taken for granted and it is assumed that it is possible to make a more or less faithful representation of the universe through the method of science. These ontological and epistemological assumptions about nature and science, respectively, have also become the commonsense view of our world. Science, in this account, is regarded as that enterprise which is concerned with providing an accurate account of the objects, processes and relationships occurring in the world of natural phenomena. The characteristics and description of the physical universe are assumed to be independent of the preferences or intentions of its observers. The unbiased gathering of data on empirical regularities observed in nature provides the evidence for formulating universal laws. Because science accepts empirical verification as the arbiter of truth, scientific knowledge is considered to be independent of subjective factors such as personal prejudice, emotional involvement and self-interest.

Though the critiques of science have led to the development of the interdisciplinary field of science studies, it is important to locate and problematize the science issue within the explanatory model of every discipline. In sociology, it is of great importance to inquire into the social factors that go into the making of the method, content and form of science. Simultaneously, efforts to 're-form' science combined with (or seen as a part of) the critiques of science, emerging from cultures or sub-cultures that have been historically displaced by modern science, can steer the process of science criticism. The outcome, at its weakest, would be a check on the dominance of science and, at its strongest, displace science from the centre-stage of knowledge production.

In order to do this, firstly, I would argue for accepting a basic tenet of sociology of knowledge that the distinction between everyday and systematized knowledge needs to be problematized. Michael Mulkay (1979) suggests that given the general vastness of the scope of sociology of knowledge, one clear and useful way of doing this would be to minimize the distinction made by Peter Berger and Thomas Luckman in their classic text (1996), between the study of popular belief/commonsense/everyday knowledge, and systematized, specialized knowledge. Such a distinction arguably reinforces the hierarchical divide between scientific knowledge/ beliefs and other knowledges/beliefs, while building a continuum loosening the boundaries between these two realms would provide the space required to develop science criticism. Secondly, the sociological study of science has been continually plagued by the truncating distinction, inherited from the philosophy of science itself, between what is known as 'the context of discovery' and 'the context of justification' of a scientific result (Reichenbach 1938). Due to this the process of discovering a result is relegated, leaving only the process of justifying the results in the realm of science. The entire context of discovery is dubbed 'social' and the context of justification as 'conceptual'. This needs to be challenged by seeing the relationship between scientific knowledge and social practices as mutually-affecting categories. Thirdly, as suggested by Mulkay, the period of transition of scientific creativity from private speculation to formal demonstration requires thorough analysis in order to understand how science uses cultural resources in research. From the private to the public, he argues, the scientist often 'smuggles in'–either as metaphor or through interpretation–her/his individual system of beliefs (ibid.). I would further argue that not only what is smuggled in as 'fingerprints', but also what is left out as 'erasures' is instructive in the sociological scrutiny of science.

This chapter is an attempt in these directions. While this approach provides us with insights into the role of the social and cultural context in the making of scientific knowledge, it is necessary to combine it with other critical approaches to science, for instance, the feminist approaches. A combination of approaches is not only productive for sociology and women's studies but also to our general understanding of science itself. Centrally, feminist theories of science require and, therefore, develop the idea of social construction of both gender and science. Science within these theories is not simply defined by 'the exigencies of logical proof and experimental verification' (Keller 1992: 25), and gender is not defined by 'biological necessity' (ibid.) alone. While science is perceived as 'the name for a set of practices and a body of knowledge delineated by a community, gender is the name for ideas about men and women defined by culture' (ibid.). Further, the feminist critique of science posits the relation between science and gender as a mutually affecting one. It is argued that not only has 'the making of men and women affected the making of science, but that the making of science has also affected the making of men and women' (Keller 1985: 4).

Though, as previously stated, developments in science studies have firmly located science within the context of culture, changing our perception of the science-culture relation, it did not directly address issues of gender. Women's studies has changed our ideas about the relation between gender and culture without necessarily addressing science (Keller 2001). With the 'maturing' of both the fields, it has become possible to look at science as socially constructed and gender as not only a real category shaping men and women but also a symbolic operator in the construction of social spaces like science. Since the 1990s some feminists have argued that science is a critical tool in reforming the canons of gender and patriarchies. Others were concerned with how science reproduces gender ideologies; while still others have used gender as a critical tool to understand scientific canons. The subsequent reclaiming or disclaiming of science in feminism is thus dependent on how science is legitimized within feminism. Harding classifies feminist responses to science in a three-fold category. She suggests that feminist empiricism claims the legacy of science, moving forward to remove androcentrism in science, Feminist standpoint theorists on the other hand would argue that feminists must bring in women's locations and experiences to bear on knowledge production in science thus producing a different science. Feminist postmodernists would push the argument even further to argue that all knowledge-claims made by science are constructed through language and theory, thereby challenging any supremacy of scientific knowledge over other truth claims. (Harding 1988). Whether posed as the 'Women's question in science' or as the 'Science question in feminism', these questions are essential and valuable in the understanding of the overall relations between gender and science (Chadha 1997; Nanda 1996).

The feminist search for a historical understanding of science has led to the critical examination of the distinctions between the Cartesian binaries of mind and matter, rationality and intuition, objectivity and relatedness, neutrality and value-imbuement that were foundational to the development of scientific rationalism in Europe. Feminists argue that these dichotomies are embedded in the practice of modern science. Arguing that in this dichotomy, what is associated with science is also associated with the masculine, 'the male-ness of the man of reason', Genevieve Lloyd (1996: 41) states that this distinction is no superficial linguistic bias but has its roots within the western philosophical tradition: rational knowledge 'has been construed as a transcending, transformation or control of natural forces; and the feminine has been associated with what rational knowledge transcends, dominates or simply leaves behind'.

The Cartesian binaries, feminists argue, reproduce gender ideologies on multiple axes. The mind/nature or reason/nature dualism is only one of a whole '*set of interrelated and mutually reinforcing dualisms which permeate western culture*' (Plumwood 1993: 42 emphasis mine) and which has implications for gender. The set of dualisms includes:

Culture/Nature
Reason/Nature
Male/Female
Mind/Body
Master/Slave
Reason/Matter (physicality)
Rationality/Animality
Reason/Emotion (nature)
Freedom/Necessity (nature)
Universal/Particular
Human/Nature (non-human)
Civilized/Primitive (nature)

Production/Reproduction (nature)
Public/Private
Subject/Object

Plumwood (43) argues that 'to read down the first side of the list of dualisms is to read a list of qualities traditionally appropriated to men and to the human, while the second side presents qualities traditionally excluded from male ideals and associated with women, the sex defined by exclusion'.

The gendering of science is described most eloquently by Keller (1985: 83): 'Having divided the world into two parts—the knower (mind) and the known (nature)—scientific ideology goes on to prescribe a very specific relation between the two. It prescribes the interactions that consummate this union, that is, which can lead to knowledge. Not only are the mind and nature assigned gender, but in characterizing scientific and objective thought as masculine, the very activity by which the knower can acquire knowledge is also gendered'. Additionally, one can argue, that knowledge-making is further gendered in the way epistemological categories are either masculinized or feminized. I would like to submit that intuition as an epistemological category—as a means of knowing—is placed on the second side of the Cartesian binary of hierarchies.

Intuition, historically has been perceived as a characteristic associated both with women and also with non-scientific systems of creative knowledge production like art and mysticism. The standard understanding of science accords very little space to the role of intuition in scientific creativity. While the 'normal' scientist in the Kuhnian sense is assumed as working in an extremely linear and rational manner, the 'exceptional' or 'genius' scientist is often characterized as possessing remarkable intuitive abilities which play a great role in their scientific abilities. In this sense, intuition is both marginalized and appropriated by the meta-narrative of scientific creativity. I would contend that from a feminist perspective, this complex status of a perceptibly 'feminine' epistemological category becomes interesting, and I aim at excavating the presence of intuition in the creative process of 'normal' science and it is instructive of how the symbolic and tacit role of gender operates in the 'masculinization of science'.

From a postcolonial location, this dimension takes further significance. Nancy Hartstock (1990) argues that marginalized sections like women and the colonized, who cease to be a subject of history and become merely the colonizer's 'other' are often placed in the same Cartesian binary, the second side. Hartstock connects this with Edward Said's well-known formulation, namely, that Orientalism is a western style of dominating, restructuring and having authority over the Orient. Hartstock points out how, in the construction of these power relations, the Orient is often feminized. It is revealing that while the colonized, along with their knowledge traditions are feminized, the modern western science they receive as colonial societies is seen as 'universal' and 'superior', a 'boon' that they have to be thankful for. And yet, narratives of scientists in these societies bear evidence to the fact that the process of inclusion in this science is conditional to many cultural erasures (Subramanian 2000). Just as across the world women in science are expected to 'harden' themselves, erasing gender identities, postcolonial subjects and cultures of knowledge-making are expected to bear the additional burden of erasing their cultural contexts. This becomes evident in most narratives of the research work of scientists in India, men or women.

With these intentions, of one, releasing the social aspects in the doing of science in order to situate science within the social, and two, simultaneously looking at the role, erasure and appropriation of a 'gendered' epistemological category like intuition in the making of what constitutes scientific knowledge, I present the map of the creative process in science based on the results of an empirical study that I conducted. The main distinguishing feature of this study was that it was qualitative and based primarily on extended in-depth interviews with scientists from the Tata Institute of Fundamental Research (TIFR) in Mumbai, a premier institute of scientific research established by Homi Bhabha, around the time that India became an independent nation (Chadha and Kamat 2002). These interviews were loosely structured and more in the nature of long conversations with the respondents which allowed for the emergence of personal narratives. Often reflective and sometimes introspective, these narratives brought forth much of the autobiographical details and personal world-views of the individual scientist. Equally valuable information, however, was also obtained from informal interactions with the respondents which happened beyond the time set for the formal interviews with them. The other method of seeking information from the respondents was through asking them to respond, in writing, to selected passages dealing with issues relevant to the sociology and methodology of science. In order to access the bigger picture or the larger context, the study was also informed by other inputs not limited to the respondents but extending to the larger community, constituting what I have called the 'received culture' of the professional scientist. These inputs included biographies of scientists, writings by scientists on issues of scientific creativity and genius, lectures by distinguished scientists visiting TIFR, in order to access both the personal and the public worlds of the scientist. The personal home pages of several scientists working in TIFR was also analysed and web pages of scientific journals and journal manuals were also studied.

In addition to the 'active' methods of collection of the empirical material, a 'passive' method of collecting information about the community was underway throughout the entire period of the study. This happened because of my role as a 'participant-observer' within the community of scientists that I was studying. In fact, I knew all my respondents personally and have interacted with them closely before and through the period of this study. This proximity to the respondents allowed me to obtain a very close look at their professional lives, their personal struggles, their passionate involvement with their subject and their stories about the 'politics' both within the Institute and in the housing complex where they resided in. I had, therefore, ample scope to discuss issues related to the methodology and philosophy of science and many of the insights that I set out in this chapter owe as much to these informal interactions which were spread out through the years of my friendship with them as they do to the more formal methods which I outline in this paper.

The choice of focussing on scientists from a single research institute was determined by the need to ensure a degree of homogeneity in the academic backgrounds among the respondents in terms of career graphs, experience and also by the rationale that the larger and general environment would be the same for most of the respondents.

Most of the existing empirical studies in the sociology of science have focussed on the study of scientists and working relations within one particular area of science like biology or experimental high energy physics. TIFR, however, has several departments dedicated to research in a broad spectrum of sciences covering mathematics, physics, computer science and biology. Close interaction with the scientists from different disciplines led me to understand that there is a rich diversity to be explored by talking to scientists from different disciplines. The education, the cultural make-up, the philosophical attitudes and the practices all showed a significant amount of variation between different groups of scientists, leading me to believe that a lot of very important empirical information would be lost if the diversity was not included. It was this realization that led to the choice of respondents from different areas of the pure sciences.

The research areas that are spanned by the selection of respondents include Computer Science, Mathematics, Theoretical Physics, Theoretical Astrophysics, Atomic and Laser Physics Experiments, Nuclear Physics Experiments, Experimental High Energy Physics and Biology so as to cover all the major areas of research in TIFR.

Moving In and Out of the Public and Private: Three Stages of the Creative Process

The creative process in science begins for the individual scientist at a social level. The idea for a research problem germinates in seminars, from preceding research papers and informal discussions with colleagues. This is largely what I call the 'public space' within the scientific community. Gradually, the 'problem' takes shape and shifts into the 'private space' of an individual scientist, that is, their own mind or within the space of a closed collaboration. Finally, the work appears in the public space as a 'result'. Thus, three very distinct phases can be identified in the creative process in science: germination of an idea, arriving at a creative breakthrough and going public.

Contrary to the notion that ideas in science emerge within the cognitive processes of individual scientists, new ideas for research emerge most often from reading existing scientific literature, from listening to seminars or through informal discussions with others. Discussion and sharing of ideas seem to be most crucial at this stage. The moment the idea takes concrete shape there seems to be a withdrawal from the public space into the private where only the individual scientist is at work alone or with collaborators. This may happen both because the private space seems to provide the best conditions for scientists to carry out their work and because once the idea has crystallized, a sense of urgency takes over. Often this urgency is driven by the anxiety that the work might be 'scooped' by a competitor. It is interesting that if in a closed collaboration, at this stage there is a further withdrawal into their individual space, making the private even more private. The moment of breakthrough, which often happens in this stage is often experienced in isolation. Once the breakthrough happens, the scientist traces the journey back into the public space of the community, first to the collaborators to whom the results are communicated. These are often partial, incomplete or even tentative. The process of reaching out to the wider scientific community then begins when the result is 'ironed out', the solution checked out. This communication is in the form of a publication.

The first and last of the three phases, I argue, are steeped, at different levels, within the 'social', and are therefore marked by the culture of science. The second phase, the most 'individuated' of the three, is where one finds signatures of the presence and role of intuition as an epistemological category.

Germination of the Idea, The Creative Breakthrough and Going Public: Voices of the Scientists

In the interviews, the respondents provided more concrete material for the first phase, which begins with the germination of the research idea, and the final phase, which involves the preparation of the final result for communication to the wider scientific community. These two phases appear to be well chartered, while the intermediate phase, where the withdrawal into the private space and the breakthrough happens, is where the least amount of descriptive insights are available. Though scientists take these breakthroughs for granted there is very little articulation about them. Whether this is because the scientist does not have the conceptual framework or vocabulary, or whether it is fraught with ambiguities at variance with the certitudes demanded of science, are questions that arise.

Over the wide spectrum of respondents ranging from mathematicians, computer scientists and theoretical physicists, who all deal with more conceptual and theoretical ideas in their work, to the more pragmatic experimental physicists, chemists or biologists, there seemed to be a remarkable unanimity in describing how a research idea germinates. Most often a scientist's research projects are the outcome of already published work and steps in an ongoing research programme. Nonetheless, almost all scientists agreed that formal and informal interactions with other scientists are a crucial component of the phase of idea germination. This happens through formal interactions in conferences, seminars or talks or through informal one-on-one interactions. Most respondents felt that reading papers and surveying the existing literature on the subject complements but cannot replace verbal interactions. Indeed, the social interactions among scientists seem to be even more important for the experimentalists, germinating ideas for research and providing solutions to technical problems related to their laboratory equipment and infrastructure.

The second phase, where a breakthrough is achieved, is clearly the most important but the respondents' description of this phase lacked the clarity that characterized their articulation of the first phase. Given a research idea, the scientist needs to have the technical know-how to tackle the problem as also commitment, which some describe as being obsessed with the problem. They all agree that these are necessary but not sufficient. As a mathematician who was interviewed put is succinctly, *'You need something extra to get to the point where a breakthrough is made. The point is you need a breakthrough and then techniques take over.'* This private moment appears to need that 'something extra' but what that is remains unspecified and vague. Most respondents insisted that there were no methodological rules that guided them in this phase and some even used words like 'irrational' to describe the muddled nature of work. One scientist felt that if he shared the notes that he kept at this stage of the work, which he referred to as his 'personal thoughts', other scientists would think he was a crank.

In the ambiguity that characterizes this phase, one thing is clear: it is a retreat into a private space. Surprisingly, this seems to be true even with the experimentalists although experimental activity takes place in collaborations, either big or small. One experimentalist talked of how he would become 'unsocial' when he had reached this stage of working on a problem. Also the responses revealed that while the withdrawal into the private space is mainly due to the engagement with the problem, it is also partly the need to keep the details of the research work confidential. This is particularly true for the experimentalists. Since the raison d'être of experimental science is the discovery of new phenomena, a good part of experimental creativity lies in having appropriate laboratory conditions for the new phenomena to be observed. This requires immense investment of time, money and effort, so an experimentalist has also much more at stake when she decides to follow a hunch and set up the experiment for it. An experimentalist said that even in the large collaborations, she tries to pursue a line of research which is marked by her individual stamp, following her own idea within the dynamics of the bigger group.

In some areas of physics and even more in chemistry and biology, the sharp distinction between theoretical and experimental research does not always hold. An atomic physicist, an experimentalist who also does theoretical work, felt that his theory allowed him to dabble in a world of imagination whereas the experiments kept him bound to reality; doing theory spurred his creativity even in the laboratory. On closer examination, he seems to be saying that it is his withdrawal into his private space that is spurring his creativity–even in the laboratory. His theoretical work allows him, not unlike the mathematician, to play freely with his imaginative and intuitive faculties. In negotiating the dual roles of experimentalist and theorist, however, he seems to discover a 'hardening' of the boundaries between the rational and the non-rational when he is at work in the laboratory, whereas the theoretical work gives him a sense that these boundaries have somehow loosened. The atomic physicist laughed when asked where he thought his research ideas originate, saying, '*The questions you are asking me, everyone wonders about these questions but nobody talks about it.*' It was a very honest admission of the silence that surrounds the issues that go beyond the notions about research laid down by the conventional method of science. He felt that since he was working in a field with an established tradition, he could make contact with other people's ideas through reading articles or talking to people to build up a knowledge base, but it did not explain where original ideas come from.

It is interesting to speculate why the creative phase in science finds such little articulation in the scientific community. This phase of scientific knowledge-production does not follow any prescribed method. As theoretical physicist David Bohm (1998: 2) says, 'For thousands of years people have been led to believe that anything and everything can be obtained if only one has the right techniques and methods. What is needed is to be aware of the ease with which the mind slips comfortably back into this age old pattern. Certain kinds of things can be achieved by techniques and formulae, but originality and creativity are not among these. The act of seeing this deeply (and not merely verbally or intellectually) is also the act in which originality and creativity can be born.' In a biographical essay on the theoretical physicist, Richard Feynman, S. Schweber (1994: 473) writes, 'Feynman learned to walk the tightrope between the psychological needs of his self and the requirements imposed by belonging to a community. He came to accept and appreciate the fact that the act of creation was for him also an act of consummate isolation'. This is a illuminating comment on how the scientist has to negotiate the space within the community to discover the private space for his own creativity.

One of the earliest works to discuss the role of intuition in the process of discovery in science is the classic work of Hadamard (1945). A mathematician himself, Hadamard attempts a psychological study of scientific creativity by studying the creative process at work in eminent mathematicians like Henri Poincaré. In fact, Poincaré, one of the most celebrated mathematicians of the twentieth century, has himself written and lectured on the subject of creativity in mathematics (Poincaré 1913). Hadamard's understanding is that invention or discovery in the sciences takes place by combining ideas. Of many combinations of ideas that the scientist may have to deal with only very few would be of real interest. Invention, according to Hadamard, is then making a choice of the right combination of ideas. The question is then: how does the creative scientist arrive at the right combination? Hadamard sees the choice being imperatively governed by scientific beauty, felt rather than formulated. Through his reading of Poincaré, Hadamard seeks to explain scientific creativity using the idea of beauty, that is, by attributing an aesthetic value to the scientific process.

While Hadamard and Poincaré locate the origin of genius in the unconscious, they have been criticized by Lewis Wolpert (1992: 63) who says, 'Intuition, as used by Poincaré and others, is no more than a convenient black box which contains the creative process but about whose workings we are ignorant'. Wolpert does not throw further light on this, except to point out that scientific intuition should not be confused with 'everyday' intuition, because he claims, 'Scientific intuition relates not to commonsense experience but to the great fund of highly specific knowledge that has been acquired'. While it is clear that scientific intuition is a special kind of intuition which presupposes considerable scientific training, it clearly goes beyond the well-laid out method of scientific discourse. As physicist Hideki Yukawa (1973: 123) says, 'Even the research worker striving to bring his creativity into play, does not know, when he feels that it has manifested itself, just how he achieved it. He has the feeling, rather, that something unexpected even to himself has taken place. Invention and discovery always have something of the nature of the unpredictable'.

In the final phase of scientific research the scientist again moves into the public sphere of her community in order to communicate the results of her research. Very often, this process starts with informal discussions with colleagues and simultaneously writing a paper that is available to the community as a 'preprint' (i.e., before it has been refereed and published by a journal).

Publishing the paper in a scientific journal, therefore, is the end result of the third phase of scientific activity. In making the passage from the private to the public space, the scientist is very conscious of the norms within the community and the manner in which he should communicate his result. This is a crucial 'make-or-break' phase. The work gains importance only if the community accepts, appreciates and takes notice by citing it; indeed, citation indices are considered very important measures of success in some areas of science. Due to the importance that the scientific community attaches to publications, there is often a great deal of pressure on scientists (especially the younger ones) to publish. One respondent, a mathematician, felt that because of the pressure to publish, she finds very little time to learn more about areas of mathematics with which she is not familiar. Further, she felt that the community does not reward this knowledge in any way so it was important to maintain a balance between production of papers and building up a knowledge base.

Presentation of the result so that it is accepted by the journal is very important. Scientific journals lay down strict norms and offer no scope for deviation. The logic of a published scientific paper is often linear, even though the results may have been arrived at by exploring more than one circuitous route. The scientist strives for clarity of thought, economy of expression and an overall clear minded and direct approach, which not only is not the reflection of the actual process of work, but also deliberately masks it. Even the language in the papers is required to be 'scientific' and liberty in the use of language is often not encouraged. Going through the author instruction manuals of journals like *Nature, Physical Review Letters (PRL), Algorithmica* and *Journal of Number Theory*, which are representative journals where the respondents publish their papers, one finds several examples of what the journal expects in terms of presentation. The American Physical Society's prestigious *PRL* even mentions that the description of '*ordinary editorial practices are not meant as an exposition of rigid rules but as an outline of usual practices*' but adds parenthetically that '*there are a few rigid rules*' (ital mine). The top-ranking journal *Nature* carries a description of how an article should have a summary separate from the main text: 'This summary contains a brief account of the background and rationale of the work, followed by a statement of the main conclusions introduced by the phrase 'Here we show' or equivalent phrase'. At another point in the manual it is pointed out that 'any discussion at the end of the text should be as succinct as possible' and also, 'Essential but specialized terms should be explained concisely but not didactically'. The *PRL* manual states that 'Any submitted comment must be cast in a collegial tone, free of polemics' while the *Journal of Number Theory* requires that the submission be in a 'clear, concise and grammatical style'. *Nature* refers authors to another web-page to find out more about writing a paper and recommends that they look up the books written on the subject. An example from such a book is: 'Writing should be as far as possible natural–that is, not worn like a Sunday suit and not too far removed from ordinary speech, but rather as if one were addressing one's departmental chairman or other high-up who was asking about one's progress' (Medawar 1979: 63). Clear, concise, succinct, non-polemical, non-didactical and deferential are some of the qualities that are valued in good scientific writing. While, on the face of it, such directions are meant for successful and effective communication, one can see how they would not be hospitable to findings that are complex, ambiguous and diffuse.

In their interviews, respondents talked about the efforts they undertake to present their results in a form that the journal will find 'acceptable'. This packaging of the result is crucial and to some extent authenticates the scientific result. Moreover, the packaging is not always purely cosmetic. For example, experimentalists have been known to present only those portions of their data which they think will make the paper acceptable. Theorists will not discuss what their models fail to explain but will only discuss the successes of the models simply to ensure publication. Respondents talked about how the same result is presented very differently when sent to a journal for publication (and thereby recorded for posterity) as compared to when it is presented in a seminar or a conference talk. In a talk, scientists would often take more liberty in emphatically presenting what is only suggestive in their results, whereas in a paper sufficient care would be taken to ensure a cautious tone while presenting something which cannot be fully substantiated. Most science journals have a process of peer review before the paper is accepted for publication. In some cases, more than one reviewer is consulted. The peer review process has again been the subject of criticism both from members of the scientific community who have had to bear the brunt of an unfair scientific review but also by sociologists of science like Resnik (1998) who have studied the ethics of the publication process.

The Construction of Genius in Science

Intimately tied up with the question of creativity is the question of genius, often associated with exceptional creativity. Work in science is often categorized in terms of creativity and productivity, the former being more pronounced in exceptional work and the latter being characteristic of routine science. I have refrained from making this very strict distinction between normal and exceptional science simply because aspects of creativity and productivity co-exist in the work of every scientist. However, often accounts of geniuses in science construct them in such a manner as to make them appear completely different from normal scientists. The difference is due to the intuitive abilities that a genius is said to possess. The difference is not, to use the language of Kuhn (1963), attributed to the convergent thinking abilities but rather to the divergent thinking capabilities of the genius. For Kuhn the tension between convergent and divergent thinking pervades all scientific work, both normal and exceptional. However, while the intuitive aspects of the work are obliterated in normal scientific work, these are resurrected in the description of the work of a genius.

It appears from the interviews that there was a notion of a scientific genius in the community though there was a range of views on what constituted genius. While most scientists had to struggle to strike a balance between being productive researchers and widening their knowledge base, it was felt that a genius was one who could do this with ease, adding the dimension of creativity to research. But genius was also viewed as the ability to see things in a different way and bring about a discontinuous change. Several respondents emphasized this ability, in some sense echoing the Kuhnian notion of a paradigm shift. But the mathematicians had other ways of deciding exceptional talent. For one mathematician, it was important to see and know a person before she can decide whether they are exceptionally talented. For her, an exceptional mind holds surprises, is not predictable and interactions with them allows her to see how these minds work. She felt that the workings of the mind are not revealed in the papers but are available only in these interactions. Another view was that a genius was one who was endowed with a vision, which was talked about as the ability to demonstrate connections between areas which apparently had nothing to do with one another. One mathematician also admitted that it was difficult to know the difference between a genius and a crank. After giving it more thought, he felt that it was essentially the lack of training that makes a person a crank. But, having said this, he found it difficult to explain why he would consider the mathematician Ramanujan to be a genius. He insisted that Ramanujan was a trained mathematician, in spite of the fact that he had very little formal education. The question that naturally came up was if we accept that Ramanujan was well trained, in spite of his lack of formal education in higher mathematics, how are we to then decide that the 'crank sending us mails' is not similarly trained and, therefore, not a crank at all. This revealed that 'geniuses' do not simply exist: they are 'named into existence' by the scientific community.

Another interesting feature of the discussion on individuals of exceptional talent was that most respondents mentioned 'non-scientific' details of their personal life. There were references to Einstein's troubled childhood and one respondent went on to conjecture a relationship between personal suffering and scientific creativity. Another talked of Feynman as a genius who did his physics while sitting in a bar but also stood in judgement calling him a 'villainous womaniser'. There was an opinion expressed that people of exceptional talent were not normal. Mathematicians mentioned a colleague, C. P. Ramanujam, who was exceptionally talented but committed suicide. This points to how a genius is constructed within the scientific community: as abnormal in personal life. Like any other community, scientists too create their own set of myths, sometimes taking liberties in fictionalizing the personal details of the heroes who have made great contributions. These myths, which form a part of the received culture of science, are not always written biographies but also survive as an oral tradition, a folklore. They have the power to transcend the barriers of the different sub-areas that science creates for itself, so it is not surprising to find experimental atomic physicists talking of theoretical physicists as their heroes. It is also interesting that most of the figures in science who are thrown up as geniuses are theoretical physicists. The exceptions that the interviews provided in this regard are also revealing: mathematicians had their own geniuses and in theoretical computer science the names of contemporaries working within India came up, characteristic of the freshness of this relatively young science. Again, a very different perspective emerged from a biologist who says, 'Biology is not driven by individuals. It is difficult for me to name some biologist as a genius because we don't make these larger-than-life persona in biology. It is funny that I can think of a couple of such names from physics but not from the biological sciences.'

In the course of making heroes, the scientific community simultaneously discards other scientists who have also contributed significantly to science. The narrative that is constructed is linear; there is no diversion from the 'main story-line'. With the scientists some of the knowledge that they dealt with is also discarded and forgotten. The result is that the scientific community, by and large, lacks a sense of history. The history that is passed on is a constructed story of a few heroes and the works and lives of several others relegated to the background or even completely obliterated.

Almost all the physicists interviewed talked of just two or three geniuses. Newton, Einstein and Feynman seemed to figure in most of these lists. It is not surprising that these names should appear in the interviews given the monumental contributions made by these scientists. What is surprising, however, is that hardly any other names were mentioned. It is even more surprising to note this when quite a few respondents talked about how even the great contributions in science are helped by all the little contributions that happen around them. This incremental view of progress in science is quite at odds with the monolithic view of the story of science that they seem to present here. Again the dichotomy in the scientist is revealed between what the received culture of science tells him and what his own experience tells him.

Knowledge-Production and Creativity in Science: A Complex Erasure of Intuition

In the discussion of process of knowledge-production in science, three phases were identified: *(i)* the genesis of the problem; *(ii)* the creative breakthrough; and *(iii)* the phase of going public with the result. The second phase, when the creative breakthrough is achieved, is the phase when a 'rational' map seems to break down. This is the phase which is the private moment of the scientist's creativity and when intuition makes its clearest appearance in the creative process. It is the breakthrough that leads to the result which is then publicly announced by means of a scientific publication. But the result that makes its appearance in the public space is shorn of aspects such as intuition and what is presented is the 'objective' result which is, at least in principle, reproducible by any other equally trained member of the scientific community. This reproducibility of scientific results is considered the hallmark of its method. As mentioned earlier, this owes a lot to the way scientific results are presented–because what is presented for the consumption of the community is what is reproducible. The proof of a theorem may be presented but not the intuitive insight which enabled the scientist to arrive at the proof. The proof is, of course, reproducible but the surprise at what led to the notion that such a proof was possible remains forever unspoken about.

This is not to say that within the scientific community there has been no attempt to address the issue of rationality versus intuition. For example, there is Hadamard's study of the intuitive process in the process of discovery in science. The eminent mathematician, Harish Chandra, said, 'I have often pondered over the roles of knowledge or experience, on the one hand, and imagination or intuition, on the other, in the process of discovery. I believe that there is a certain fundamental conflict between the two, and knowledge, by advocating caution, tends to inhibit the flight of imagination. Therefore, a certain naïveté, unburdened by conventional wisdom, can sometimes be a positive asset' (quoted in Langlands 1985: 235). However, these views remain on the margins even when they are expressed by eminent scientists like Harish Chandra.[1] This is because the hold of rationality and objectivity on scientific epistemology is so great that it is difficult to break free of the shackles they impose.

In a feminist reading of this 'hardening' of the creative process, where the privileging of rationality and objectivity and the marginalization of intuition and imagination occurs, the hardening is indicative of the gendering of science. This epistemology is, as has been realized through recent feminist discourses, a masculinist one. To quote Simmel: 'The requirements of correctness in practical judgements and objectivity in theoretical knowledge belong as it were in their form and their claims to humanity in general, but in their actual historical configuration they are masculine throughout. Supposing that we describe these things, viewed as absolute ideas, by the single word 'objective', we then find that in the history of our race the equation objective = masculine is a valid one' (quoted in Keller 1985: 75). In the feminist interpretation, therefore, the marginalization of intuition is a clear indication of the gendering of science. It is in this process of gendering that the subjective aspects of the creative process–the intuition and imagination that Harish Chandra refers to–are completely obliterated. In the final presentation, this erasure of the 'subjective' is so complete that every reference to the human observer is removed. Referring to this as 'context-stripping', Ruth Hubbard (1989: 125) writes, 'Natural scientists describe their activities as though they existed in a vacuum. The way language is used in scientific writing reinforces this illusion because it implicitly denies the relevance of time, place, social context, authorship and personal responsibility. We can see this process of context-stripping clearly in the manner the intuitive aspects are obliterated from the creative process in science.

While this marginalization of intuition takes place in normal science, the site where it makes a re-appearance publicly is in the construction of scientific genius. Biographies and essays on geniuses in science are replete with stories of the unusual minds they possessed, how their thought processes were completely 'different' from the set processes encountered in routine science and, very often, about how it was so difficult for the ordinary scientist to reproduce the method of the genius. A mind finely tuned to arrive upon intuitive insights remains an enigma for science. In turn, the community celebrates this mind as a genius. In this twist to the process of gendering in science, the marginalized feminine aspects are recovered not in a female form but in the form of a male demi-god, the genius.

Unlike the other erasures discussed, the erasure of intuition from the creative process in science is complex because while it is marginalized, in normal science, it is also appropriated, in the construction of genius. Unlike the other erasures, it is not a mere re-surfacing that is occurring in the case of intuition because the intuitive elements of the creative process cannot be wished away easily. In other words, it is hardly possible to discuss a scientific genius without discussing intuition even though the rational discourse of science attempts to marginalize it.

Note

1 Harish Chandra was an eminent mathematician who made fundamental contributions to group theory. He was professor of Mathematics, Institute of Advanced Study in Princeton. He obtained a Ph.D. in theoretical physics in Cambridge before making the decision to quit physics and take up mathematics instead.

References

Berger, P.L. and Luckmann, T. 1996. *The Social Construction of Reality: A Treatise in the Sociology of Knowledge*. Garden City, NY: Anchor Books.

Bloor, D. 1976. *Knowledge and Social Imagery*. London: Routledge and Kegan Paul.

Bohm, D. 1998. *On Creativity*. London: Routledge.

Chadha, G., and Kamat, V. 2002. 'A Scientist's Patronage of Art: Homi Bhabha's Art Collection', *Humanscape* 9: viii.

———. 1997. 'Sokal's Hoax and Tensions in the Scientific Left', *Economic and Political Weekly* (henceforth *EPW*) 32, 35: 2194–96.

Hadamard, J. 1945. *The Psychology of Invention in the Mathematical Field.* Princeton, NJ: Princeton University Press.

Harding, S. 1988. *Feminism Science, and the Anti-Enlightenment Critiques.* In *Feminism/Postmodernism* edited by Linda Nicholson. New York: Routledge and Kegan Paul, 83–106.

———. 1986. *The Science Question in Feminism.* Milton Keynes: Open University Press.

Harstock, N. 1990. 'Foucault on Power'. In *Feminism/Postmodernism*, edited by L. Nicholson. New York: Routledge: 157–75.

Hubbard, R. 1989. 'Science, Facts and Feminism' in *Feminism and Science*, edited by N. Tuana. Bloomington, IN: Indiana University Press: 119–31.

Keller, E. Fox. 2001 'Gender and Science: An Update'. In *Woman, Science and Technology: A Reader in Feminist Science Studies*, edited by Mary Wyler et al. New York: Routledge.

———. 1992 *Secrets of Life, Secrets of Death.* New York: Routledge.

———. 1985. *Reflections on Gender and Science.* New Haven, CT: Yale University Press.

Kuhn, T.S. 1963. *The Essential Tension: Selected Studies in Scientific Tradition and Change.* Chicago: University of Chicago Press.

Langlands, R.P. 1985. *Biographical Memoirs of Fellows of the Royal Society of London*, no. 31, London: Royal Society.

Lloyd, G. 1996. 'Reason, Science and the Domination of Matter'. In *Feminism and Science*, edited by E. Fox Keller and H. E. Longino. Oxford: Oxford University Press: 41–53.

Mannheim, K. 1936. *Ideology and Utopia.* Fort Worth: Harcourt, Brace and World.

Medawar, P. 1979. *Advice to a Young Scientist.* New York: Basic Books.

Mulkay, M. 1979. *Science and the Sociology of Knowledge.* Aldershot: Gregg Revivals.

Nanda, M. 1996. Science Question in Post-Colonial Feminism, *EPW* 31, 16: WS1–WS8.

Plumwood, V. 1993. *Feminism and the Mastery of Nature.* London: Routledge.

Poincaré, H. 1913. 'Mathematical Creation', in *The Foundations of Science*, translated by B. G. Halsted. New York: The Science Press.

Reichenbach, H. 1938. *Experience and Prediction*. Chicago: University of Chicago Press.

Resnik, D.B. 1998. *The Ethics of Science*. London: Routledge.

Schweber, J. 1994. *QED and the Men Who Made It*. Princeton: Princeton University Press.

Subramanian, B. 2000. 'Snow Brown and the Seven Detergents: A Meta-Narrative on Science and the Scientific Method', *Women's Studies Quarterly* 28, 1 and 2.

Wolpert, L. 1992. *The Unnatural Nature of Science*. London: Faber and Faber.

Yukawa, H. 1973. *Creativity and Intuition: A Physicist Looks at East and West*. Tokyo and New York: Kodansha.

Ziman, J. 2000. *An Introduction to Science Studies: The Philosophical and Social Aspects of Science and Technology*. Cambridge: Cambridge University Press.

23

Science in Architecture and Architecture in Science: A Conversation with Neera Adarkar

Gita Chadha and Unnati Tripathi

Architecture is a discipline constructed both as a science of building and the art of making spaces. We were keen to unravel the underlying gender dimensions in the discourse and practice of architecture. Since this is an under-researched area in science studies, we decided to explore some questions in this interview. We hope to open up questions for further enquiry.

Neera Adarkar, a Mumbai-based architect, urban researcher and activist, is a partner in the firm Adarkar Associates. She has also worked closely with the Textile Workers' Union in Mumbai and has been involved in the struggles of the citizens' coalition, the Mumbai Peoples' Action Committee. She is a co-founder member of Women Architects Forum. In the early 1990s, the forum provided the impetus for women architects to revisit their profession and view the built environment from a gender perspective. Neera graduated from the J.J. College of Architecture, Mumbai and did her post-graduation from the Indian Institute of Technology, Mumbai. She has been visiting faculty at several Mumbai colleges and has given talks at different institutions in India and abroad. She has published many essays, and co-authored (with Meena Menon) One Hundred Years, One Hundred Voices: The Millworkers of Girgaon: An Oral History (2004). *Recently she edited* The Chawls of Mumbai: Galleries of Life (2011).

Neera was interviewed for the present book by Gita Chadha; the interview tapes were transcribed by Unnati Tripathi and this article was edited and compiled by Sumi Krishna. The interview is followed by an extract from 'Designing a Women's Special Train: Negotiating Spaces in Architectural Pedagogy', in which Neera describes her experience of initiating a critical social and gendered architectural pedagogy.[1]

I

The Interview

GC: *Do built spaces reflect and reproduce cultural and ideological assumptions? In style, structure, access to spaces? Consequently, do built spaces also reproduce social hierarchies of class and gender in tacit ways or in an obvious manner?*

NA: Built spaces are the physical representation of social relations. Architecture of a culture reflects social structures. This is not easy to perceive because all of us are taught to see a place only in terms of its outer facade. We need to look at how the space is ordered rather than how it is presented. It is in the ordering of space that social hierarchies of class, caste and gender are reproduced. This could be visible in the imagery used or in the way boundaries are set. For instance, some built spaces have separate entries for men and women; for the masters and servants; for upper caste and lower caste. More recent and modern structures like universities and colleges in some parts of the world have separate entrances for men and women. Very often specific locations within a built space are associated with or demarcated for women. These architectural gestures reflect gender and social hierarchies. Like the inner courtyard in traditional Indian homes which was the only space where women could enjoy the 'outside' by enjoying a piece of the sky while still remaining inside the house. Otherwise their entry into the outside public space was systematically controlled. They were not allowed to sit on the *otla* of the house which faced the street outside the house; this space was only available to men. You can then see the internal courtyard as a space resulting out of negotiations within the patriarchal structure—a semi-open space out of the public gaze. So, yes, architecture of spaces can and do reproduce cultural and ideological hierarchies.

GC: *Yes, these 'inner spaces' were the spaces from where women could see a piece of the sky. But how does this happen in terms of architectural style? Would you say architectural styles reflect the location of the architect or are they objective?*

NA: Firstly, what do we mean by style? Do we see it within a historical, cultural, technological context, within a religion or regional discourse or simply as a matter of form? Reference to style is more often than not, simply a reference to an articulation on the structure's facade. In India, we speak of a Colonial, a Hindu, an Islamic style of architecture. Now, we also speak of a global style. Interestingly, none of these styles can be unmixed categories. There is always a taking forward of a technology of construction especially within a certain regional and climatical context. Often we see a blending and fusing of styles. Some elements become symbolic of a certain style over a period of time like the arches and domes which have become symbolic of 'royal' or majestic culture. In contemporary architecture of our cities, such elements are used as symbols of luxury because people do connect them with earlier glorious era. So, the notion of style is quite limited and should be contextualized within the variables.

Moreover, style should refer to both form and the spaces within; but unfortunately when we look at the building only in terms of its facade and not its interiors, we ignore the experiential aspect within its built spaces. If we consciously do that, we will not escape its conscious ordering of functions. I would say that no style is a neutral style; no space is a neutral space. But this assumption was not questioned until the built environments like other social and cultural environments were also assessed/examined/looked at from gender perspective.

GC: *Why is that so?*

NA: There is a certain romantic glamour still associated with the architectural profession, because of its association with the 'aesthetics' of building. It is also because there is a perceived divide between art and science and that aesthetics in architecture can be seen as independent of social sciences. This is reflected in the pedagogy too. The architect is supposed to be the 'master' of the construction industry. This has given a mock air of superiority to the architects. Yet, the Issues of material, structural engineering and construction, are considered to be the 'objective' way to approach built structures. It is taken for granted that with this status, comes a neutral (and noble) approach. This is completely incorrect, as we all know. But the architects actually think that they can conceive the needs and aspirations of all the users of the spaces—irrespective of their class, gender and age—they are building for and therefore can decide what shape the built environment should take.

This results in a wide gap between their assumptions and the social reality (especially the reality of marginalized sections with which they do not identify/or empathize with) before any planning decisions are undertaken because what is planned and built now, will influence the ways of life of women for many years to come. For example, planners tend to assume a woman's role only as a 'reproducer', ignoring completely her role as a 'producer' (although the substantial contribution made by women is now recognized by economists world over).

GC: *It is apparent that, like in many other professions, women are underrepresented in architecture too. But would you say that architecture, as a discipline, is gendered in foundational ways?*

NA: Qualifying for an education in architecture in India is not easy. First of all the courses in architecture are expensive. These are not easily available choices for women in patriarchies. Yet in the cities you will find as a rule, 50 per cent of the students are female students. Being an architect is very demanding. Throughout the course we are trained to do very challenging work. As a lot of this is team work, schools are open till late at night, male and female students work together. Women students learn to rise to each challenge. We see large numbers of women working as architects in architectural firms, although not many are given an opportunity to handle building sites. It is perceived that they cannot attend to work on the building sites. It is not only because of the assumption that they are not physically capable, but because of an inequitable assumption that women are incapable of supervising structural engineering work. Fortunately this reality it now changing fast because *(a)* in most of the offices there is an equal ratio of employed female and male architects; and *(b)* as more women are into running their independent practice a large number of their women assistants get an opportunity to manage the work sites and they are very efficient. The parents of women employees are also very open to their travelling, unlike, say, a decade ago. But then the number of women who have independent practice is still very low. The reasons are like any other profession. The patriarchal pressures, and the fears of double and triple burdens still dissuade them to venture out on their own. The highly messy character of the real estate business can also be a repelling factor. As Ellen Perry Berkeley (Berkeley and Mcquaid 1989), a feminist critic, comments, 'The real problem for a thoughtful woman is not whether she is accepted into the profession but whether she wants to be accepted into the profession as it is now'.

GC: *What are the biases faced by women who are in their own practice?*
NA: There is a kind of unjust assessment about women architects that they (should) choose less stressful, 'soft' fields like Interior design (as against building design) or Heritage conservation. Firstly, both these options are not necessarily easier fields, and secondly, women who practise in these fields might have chosen them because of their interest and not because they are soft. At the same time when starting a fresh practice, men too find it is easier to get interior design

projects than building projects. But they don't get branded as only interior designers like women. Even today after so many years of practice, I do come across people who raise their eyebrows in appreciation to find my involvement in the building design. Many women architects who own their own firms, therefore, consciously consider to reduce their association with the interior design practice, if they can afford this financially, in order to challenge such stereotypes. It is true, however, that women architects face various kinds of biases quite similar to that in other professions.

GC: *What about the experience of women in the field of architecture in India? Are women architects in India conscious of the gender biases in architecture? Will it make a difference to get more women into architecture?*

NA: Well, women architects, unlike women lawyers, or even media women, have not yet delved into the gendered aspects of their own discipline. They are still unaware of their own potential to examine a built space for its gendering. There are two main reasons. Architectural pedagogy/syllabus completely ignores the significance of the existing urban context and its connection with social science and humanities. Like there is a divide between art and science/technology there is an deplorable apathy on the intrinsic connection between architecture and the study of the society at large, leave aside gender studies. So the gender awareness of women architects revolves around personal experiences common to all the working women in any the other field. Seeking more representation or gaining visibility in the field are definitely important issues but in the three or four conferences held of women architects, we have observed that any discussion about feminist critique with respect to the structure and design of a built space intersecting with the gender hierarchies is not found to be very engaging.

GC: *Some women in the natural sciences too feel that feminist challenges to the foundations of their fields are too radical.*

NA: Yes, most women architects do not want to look at the built spaces differently. Most women architects cannot come out of the mould of 'objectivity'. In trying to compete with men to be successful in the profession, they have not thought about the possibilities of connecting their own specific spatial experiences and examine if they are different from that of the men. This calls for some awareness of a feminist approach to create a broad framework and to see how this is attempted in the other disciplines. Again our pedagogy does not in any way introduce Women's Studies as a course or even as an elective. It would be interesting to expose the students to the other disciplines which have undergone such revisiting through a gender lens. This understanding arose in some of us here, because of the feminist sensibilities generated by the women's movement and the work done in the other disciplines like history and law. I guess it is the same with women in other 'sciences'.

GC: *What does it mean to look at design differently?*

NA: Some of my earlier studies in the MHADA colonies and slums of Mumbai showed me that the built environment affects men and women differently and there is no such thing as a neutral physical space. As I said earlier the planners, policy-makers, architects are products of the same educational system that makes them believe that as decision-makers they are non-biased; that the built environments they create are neutral, and therefore their perceptions about the needs and aspiration of the society for which they create spaces are unchallengeable. However the assumptions made by the architects about women are not only out of date but are out of reality with women's multiple roles in the society. They are rigid, confining women to the stereotypical roles in private and public spheres. The ordering of physical space within a built environment then reflects both, the flawed perceptions as well as intended functions within patriarchy.

But then the built space is generally viewed only as an outer form, or rather the value of the building is seen only in the facade. Further, the built form is judged by its monumentality. This aspect of 'scale' in a building is very important when you see 'differently'. The monumentality of the facade or the large inside spaces in the public buildings are awe inspiring and at the same time could be intimidating for women to respond to them without feeling inhibited in their movements. So it is important to understand how intended function of a built environment could be gendered before we design or respond to it.

GC: *Not just function, but intended function...*
NA: Yes.

GC: *Neera, could we now talk about the architecture of scientific institutions in India and what they reflect? Would you agree that architectural spaces of scientific institutions play out the vision of the founders? Like the architectural style and design of Bose Bigyan Mandir built by Jagadis Chandra. Bose reflected his vision of synthesising western science with eastern mysticism.*[2]

NA: Every building should reflect some vision, small or big. A building can reflect a lack of vision but no built space can be said to be neutral. Reflecting any vision especially a synthesis of science and mysticism in any art form is a complex process, very challenging. Especially an attempt to synthesize in architecture by combining visual elements from two different cultures directly becomes a simplistic, unsophisticated approach. Secondly, a set of architectural features associated with a certain period, culture, class, religion, region provides significant context. If you use some of the factors without their relevant context to fulfil a certain specific but an abstract objective, the failure is inevitable. I am quite wary of categorizing the styles on the basis of religion. Because there are lot of overlaps between the components mentioned above. The ruling dynasties who funded monumental structures to legitimize their power, belonged to different religions but the religious identities of the real builders-skilled masons, artisans, and so on, was not the defining factor when it came to construction.

'Styles' not only drew the inspiration from religion, a lot depended on the regional climate, local materials, and so on. But even if we go by the stereotyped elements associated with Hindu/Muslim styles, apart from using rather obvious Hindu symbols within the structure and design of the Bose Bigyan Mandir, there is no subtlety. The Bose Institute uses only symbols/imagery that supposedly represent some historical period. Arches, brackets, sloping chajjas, terrace jalis used in so-called Mughal architecture. They can be perceived differently on the basis of gender, class and caste too. As I said earlier, some features of a style persist over time and are often used to connect with an earlier era. The use of water bodies and courtyards are anyway interesting and useful features in any public building and should be appreciated in Bigyan Mandir irrespective of its religious context. I would think that Bose had a limited objective when he used Hindu and Islamic elements of architecture. He perceived this as being more 'Indian' as opposed to the colonial architecture based on the import of Gothic or Greek elements of his times. This surely reflected his brand of nationalism.

GC: *Science is often seen as best pursued in isolation, away from the 'social world'—more as an asocial activity. Science institutions are often located in isolated spaces—for example, the Tata Institute for Fundamental Research is located at the southern most end of Mumbai, on heavily secured Indian Navy property. Can you see any of this reflected in and constructed through the architecture of TIFR?*
NA: I see in the design a deliberated attempt to keep the entire community of scientists exclusive, away from the society. But equally important is that in the ordering of the space within the structure which is in a linear fashion. To me the intention seems to keep individual scientists separated at work. The long tunnel-like impersonal corridors with shut rooms on either side promotes a heightened sense of 'the creative genius' at work, independently. This cool long narrow space evokes an insecure/vulnerable feeling even within such a secured complex. There are very few shared

spaces which exude some warmth, and where people can relax or discuss with ease. This does not give a sense of a campus where interaction is expected. Indeed, the hostel is also constructed similarly. All of it appears much sanitised, clinical. It is extremely formal and impersonal. But this was the intended function, right?

GC: *TIFR, which has often been described as a temple of Indian modernism, arguably reflects the vision of its founder, Homi Bhabha, who took a personal interest in the architecture of the institute. He came from an elite westernised background and shared Nehru's vision of science and development.*

NA: Unlike Bose, Bhabha and Nehru both were looking westwards for marching forward. That would explain the choice to use the Bauhaus style of European architecture for the building. It is minimalist, stark and non-decorative. In that sense I do appreciate the building.

GC: *The Bauhaus style is a modernist one—linear, angular—can we say this might be considered to be a more masculinist style in and of architecture and alien to India?*

NA: Well, if the lack of a non-linear design, the scarcity of shared spaces, the affirmation of a certain isolationism and atomism means masculinity, it could be said that the design is masculinist. But I think this is again a kind of simplification. We should be careful of such dichotomies as we are of other patriachal dichotomies which are also hierarchical. Linearity per say cannot be associated with masculinity and curvatures cannot be linked to feminity. In fact such divisions are used to ridicule our search for feminist architecture. Parameters of feminist architecture would be very complex. Its aesthetics will vary with different regions and cultures. But surely we can identify certain basic parameters used in other disciplines to see they are represented in the built environment and how to counter them with through the medium of architecture. Like challenging the dichotomies of private and public, division of labour, hierarchies within patriarchal family structures, equitable distribution of resources, and so on. Sensibilities towards these issues can be represented in planning and design. At the same the design of spaces should also the reflect the future aspirations of women users. The gender perspective demands that it is imperative to understand the role of women in the society of our vision—a society based on equity of status and resources. Therefore the solutions need to be explored at two levels, immediate and long term that are geared towards the long-term structural changes in the power relationships.

As men who deeply identified with the western Enlightenment, both Bhabha and Nehru shared the gender ideology of that movement, something which has been critiqued by feminist historians. For both, the idea was to create a temple of modernism where the new gods of India, a large fleet of scientists, would reside. The scale of the building is huge and alienating; it inspires a sense of awe. Unlike other public buildings of similar scales, where large numbers are expected to work, the small numbers at the TIFR give it a sense of the exclusive, the special. The design clearly sends the message that this is not a space for ordinary men. The horizontal ivory tower is intended to inspire a sense of intimidation. This is quite intended.

GC: *Yes, a lot of visitors to the institution do experience that... but do we see any hybridity here, a hybridity characteristic of a postcolonial culture like ours which combines early civilizational modes with the newer colonial modes?*

The fourth floor of TIFR, where the offices of professors were located, did not have a women's toilet for a long time—the assumption being that either women won't make it to professorial levels or that even if they did they could spare time to use the loo on other floors unlike the male scientists. Further, women in laboratories have often said that machines, electric switchboards, and so on, are not designed to suit women and increase their dependence on men staff—making them disadvantaged in the eyes of male scientists.

N.A. The issue of absence, inadequacy and ill-equipped toilets for women is very common in public buildings. Mumbai's Development Control Regulations makes it mandatory to assume the number of women employees to be equal to that of the men employees while calculating the minimum number of toilet seats. But the issues of maintenance, provision of special amenities in the toilets, locations, still exist but they can be rectified—either at the design stage or even later. Similarly changes in the switchboard, and so on, should be definitely demanded by the women employees. This apathy is seen in ignoring the special needs of the senior citizens, children too.

II

'Designing a Women's Special Train: Negotiating Spaces in Architectural Pedagogy'

Conceptual Framework

'There are feminist ways of looking at and making architecture, but these are based on a certain approach that stems initially from an understanding that our surroundings are not neutral, that there is a relationship between the "content" of architecture and our capitalist and sexist social structure'(Boys 1991).

At the base of architectural pedagogy is an assumption that the architect/planner is at the helm of the profession, which in turn, is seen as neutral and non-biased. Planners, policy makers, architects are all products of the same educational system that makes them believe that as decision makers they are neutral and non-biased and the environments that they build are gender neutral. Yet it is now well known that the built environment affects men and women differently and the culture of planning and building has strong links with the politics of space. 'There is no such thing as physical plan that is "neutral" any more than a neutral technology because both are developed by men from the male point of view' (Hosken 1987).

Planners' perceptions about women as the users of the built spaces show ignorance about the complexity of women's role in the productive and reproductive spheres, and about the divisions of private and public spheres that should be the basis for any discussion on the gendering of the built environment. 'The concept of Public and Private, which has been investigated in the other disciplines, is integral to the analysis of the built environment as well. Public/ private is one of the hierarchical dichotomies. Dualities such as masculine/feminine, inside/outside, rationality/intuition, culture/ nature, spirit/flesh are not accorded equality but rather one term attains pre-eminence over the other, relegating it to an inferior status. This hierarchical system of values, which allows one group of people to dominate the other, is understood to be the product of patriarchal thought. The feminist analysis of the hierarchical value and power systems attached to the public/private sphere can be used to form a feminist critique of architectural theory' (Fowler 1984).

Any built environment ranging from a shelter to an urban space which contains within itself issues of gender, class and culture. It is imperative for planners to first understand and accept the reality that their perceptions are determined by their class and gender backgrounds, because this results in a wide gap between their assumptions and the social reality (especially the reality of marginalized sections with which they do not identify/ or empathize). And this must come first before any planning decisions are undertaken because what is planned and built now will influence the ways of life of women for many years to come. For example, as mentioned before, planners tend to assume a woman's role only as a 'reproducer', ignoring completely her role as a 'producer'.

Architectural Pedagogy

The J. J. School of Arts was the first school of architecture started in Bombay in 1922 and for a long time remained the only school of architecture in India. However, the objective of the colonial rulers was not to train students to become planners or designers but to be competent draftsmen in the service of colonial architects or to serve in the Public Works Department. Stress was given on the drawing and engineering skills, on the building services, surveying, valuation and similar fields. Ironically, till the Council of Architecture was formed in the 1970s the Architect's License was called as the Surveyor's License and could be obtained only on the criteria of the knowledge of the building regulations set by the Municipal Authority. It is not surprising that the main admission criterion for aspirants for a majority of Indian architectural schools have been their performance in Physics, Chemistry and Mathematics in the Higher Secondary School Certificate.

It took decades to shift the focus of architectural pedagogy from developing drawing skill to designing skills. Being under the control of the Technical Education Board, the understanding of design activity was limited to the physical ordering of the space based on the function on one hand, and 'façade decoration' on the other. Later institutions like Centre for Environmental Planning and Architecture (CEPT, Ahmedabad), which came under influence of the modernist masters and especially their work in India, recognized 'space' as the essence of architecture but failed to emphasise the socio-economic context, thereby underplaying the relationship between architecture and urbanism: 'The "Abstract Universalist", in denying culture and context, denied the relationship between Architecture and Urbanism. Architecture became largely the issue of the isolated or pristine objects and man' (Varkey 2000). As a result, the teaching of architectural history revolved around facts and styles under various periods without studying the social processes behind them.

Schools of architecture started in the 1980s and 1990s, made deliberate efforts to incorporate other art disciplines into the course but it remained a peripheral level rather than becoming truly interdisciplinary. Besides, these efforts were individually promoted and not always part of the curriculum.

The basic criticism about the way architecture is taught or practised stems from the fact that it lays emphasis on the built forms and scaled spaces and does not see it as a social activity as well. The building was like an image to be seen from outside, architect was seen as an image-maker and architecture as a glamorous profession. An intellectual/ideological debate was not on the agenda of architects who graduated from most schools of architecture, nor did they care to examine their own perceptions about the society for which they building.

The 'PWD' mindset of the architects infected the building culture in most of the cities in India,[3] since Sir J. J. School of Architecture was the only school in the country for a long time, their students occupied the highest posts in the Public Works Department. This resulted in a genre of architecture associated with the government bodies (Menon 1997). Till date, the profession is dominated by a culture that revolves around the skills and strategies of utilizing maximum Floor Space Index (FSI- permissible built up area on the given plot of land) and satisfying the building regulations set by the government on one hand, and by the political and social sensibilities of the urban elite and the middle class on the other.

In Indian universities the culture of intellectual dialogue among disciplines like sociology, history, literature was gradually evolving. However, these academic debates and dialogue about modernism and traditions failed to make an impact on the discipline of architecture, a situation that has changed only marginally even today. At the same time the political process for creating a climate conducive for evaluation of the architectural products and processes from a critical social perspective was absent (Adarkar 2004). The reality is that the architectural curriculum even today lacks a holistic approach. The faculty of architecture is predominantly drawn from the same academic culture, which has remained impervious to the critical issues that have been thrown up outside the architectural disciplines. Design projects in the studio respond to market-oriented demands. Postmodernism is seen as a style that has become a saleable commodity in the real estate market. Globalization is not looked at with a critical point of view of the third world economy; it is limited to the outer 'image' of the buildings that becomes a symbol for the 'multinational' global image.

An architecture school's interest in a particular design is influenced to an extent by issues being focussed upon by the media or their association with well-known architects, or is dependent on the market value at that point in time. For example, let us look at the slum redevelopment schemes or the development of the mill lands in the working class neighbourhood of Mumbai. Both became hot topics for design, town planning but mainly as a platform to display design skills. For the last five years in almost all the schools of architecture in Mumbai, the development of mill land has been part of the curriculum, predominantly to display the designing skills of the students. When the slum redevelopment policy became market-oriented builders getting double the FSI for market sale, and many reputed architects and builders helped to publicize the scheme thereby influencing faculty members and students in the selection of topics for architecture design and thesis respectively, but without discussing the ideological issues that are crucial for urban planning, such as displacement, 'encroachment', 'unauthorized' status of certain constructions. Similarly, when swanky skyscrapers and plush commercial houses were built on the mill lands, the mill lands became hot topics for design, town planning and for writing theses. In none of the subjects did the approach go beyond the built forms to seriously address the economic and cultural dimensions or to explore the politics of urban space.

In the construction of their narratives for the design studio projects, women protagonists are conspicuous by their absence. As a member of the design faculty, I had once asked my students to design a 'house for a home-based professional'. To my surprise, the 'home-based professional' was assumed to be the man of the household by 99 per cent of the students! I have observed that any attempt to discuss gender issues with the students requires a very careful approach as the response could range from empathy to a backlash from them and also from faculty members. As Dr. Maithreyi Krishnaraj (1991/1992) writes 'one does not expect a necessarily pro-marginalized/pro-poor ideological thrust but an impassioned discussion or diverse point of views should be brought forward consciously to let the student discuss, debate, evaluate and adopt.'

I have observed that in recent years out of the 600 thesis topics, only 18 cover social institutions; out of which four are women's institutions. These include the Rehabilitation Centre for Women in Distress, Women's Development Centre, and a Girls Military

School (all from the Rizvi College). A student from the Academy of Architecture had a research thesis on 'Gender in Architecture'.

Interventions

The curriculum even today fails to take note of the important role of social sciences in the study of the built environment. We need to find ways and means to infuse these in our teaching, by inducing the students and the faculty to take a critical look at social reality, particularly the reality of women as consumers, creators and critics of the built environments.

Fifty per cent of the total users are women as are half of the total number of students enrolled in the graduate colleges of India. It is important that they address the relevant issues in their thesis topics, in the electives, in the architectural design programmes, and in special workshops. Some of the issues that need to be addressed while examining the gendering of built spaces are: how does the (man-made) built environment: the house, the community and the city at large, affect men and women differently? Are the spatial needs of women different from that of men? Do these needs cut across classes and cultures? What is the critical framework for reviewing the existing planning norms/policies? What does it mean to create/ plan a gender-biased environment? How can these concerns reflect in planning policy? What is the role of women architects/planners? Will women designers express differently than their men counterparts? We need to create a an awareness among students about the status of women in the family and in the structure of the society, to look at how gender is constructed, what is the women's contribution in the economy (in terms of subsistence as a housewife and as a participant in the wage labour market); how the division of labour is based on gender and how the power relationship in the household and outside dictate its priorities and needs. Further we need to look at how those needs, including the spatial needs of women and other marginalized sections, get subsumed under those of more powerful sections of society. This can be introduced in the regular design programmes or specially formulated design programmes. For example, it is possible to examine town planning studies while addressing the needs of society that are studied for the planning of the mass housing and amenities in the higher classes. As a visiting faculty for the first year design teaching team, I have looked for opportunities/spaces where I can, both directly and indirectly, try to expose students to the gendered character of the built spaces and add a gender dimension to the design programme by starting with small design projects like a crèche, and by an exercise that makes the students analyse their own homes.

While experience shows that although there is an overwhelming majority of women students in the different schools of architecture all over India, it does not necessarily mean that they will develop a pro-women perspective. The women students (as well as women professionals) share with their male colleagues, the same urban middle-class privileges in their upbringing, which allows them an access to the professional courses but which also keeps firmly and unquestioningly in the grip of patriarchy. While the contribution made by the women's movement in creating a new awareness in society is recognized in the field of social sciences and humanities, the image of the women's movement in the minds of the students and the faculty of architecture is inadequate and biased. It results in an approach that is casual at its best and hostile at its worst.

The role of architecture has been narrowly defined by an architectural pedagogy which was articulated at an earlier colonial stage and has not been sufficiently re-examined. It has failed to create awareness towards the potential of the discipline to address the social and economic issues that play a crucial role in the determining the class and gender character of the culture of building (Adarkar 2004).

Since many of us involved in the Women Architect Forum (WAF)[4] were also connected with the various schools of architecture in the capacity of either visiting or full-time faculty, we hoped to make some meaningful interventions in the established system of architectural education. We planned to design gender studies curricula especially for architecture in collaboration with the scholars from the discipline of social sciences. In retrospect this seems like an over ambitious and unrealistic agenda as it aims for the inclusion of the entire manifesto of feminism that has not yet been accepted among the social sciences in most universities.

The immediate interventions possible in the current scenario of architectural pedagogy could be at the three levels.

(a) The interventions should be by negotiating spaces wherever possible. It may be needed to study the existing curriculum and locate the sites for negotiation. Although curriculum changes are a long-term agenda, the initial interventions could be started that would eventually influence the Board of Studies.

(b) Innovative ways of intervening in the present system will need effort to pursue the strategies to integrate our concerns in the existing setup. Some of the interventions could be in the following areas: in the selection of thesis topics and of the electives; in the programmes of architectural design studios. Each project, each programme can be looked at from a broader gender perspective to enrich its scope. A lot of empirical studies could be promoted through the theses, which I see as a more accessible site for introducing and studying different perspectives. Empirical studies can add facts and insights into the discipline. The students should be encouraged to get exposed to the research work done in the other disciplines and then learn to theorize on their experiences.

(c) Short-term multidisciplinary appraisal workshops can be offered to various colleges as elective programme. It could involve the faculty though the starting point could be the students. Women's Studies academicians and artists from different disciplines–visual and performing–can be involved in designing these special capsules.

It is, of course, too early to imagine what form and appearance buildings and cities designed from a feminist perspective would take, and how they would function. As Pauline Fowler (1984) says, 'We are dealing with an unknown iconography, and the most we can do at the moment is experiment and discuss the many questions in our minds, including that of "a women's architecture"'.

Notes

1. This extract is reprinted from Adarkar (2007) with permission.
2. See Madhumita Mazumdar, Chapter 21, this volume.
3. Public Works Department (PWD), the government agency dominated by the civil engineers who are in charge of most of the building activities in the cities.
4. In 1990, a Women Architect Forum was formed after the Indian Institute of Architects in Mumbai invited some of us to speak on the occasion of Women's Day on 8 March 1990. The first meet of WAF was in Ahmedabad, which provided a platform for sharing experiences as women professionals. In 1992 a seminar to explore the relationship of women and the built world was arranged in Bombay. WAF in Bombay also designed and conducted 'refresher' courses for the women architects who had lost touch with the profession due to domestic responsibilities and wanted to return to it. This activity received a good response, and provided an opportunity to introduce gender issues in the course.

References

Adarkar, Neera. 2007. 'Designing a Women's Special Train'. In *Gender and the Environment in India*, edited by Madhavi Desai. New Delhi: Zubaan, an Imprint of Kali for Women: 241–59.

———. 2003. 'Gendering of the Culture of Building: A Case Study of Mumbai', *Economic and Political Weekly* 38, 43 (25 October): 4525.

Berkeley, Ellen Perry, and Matilda Mcquaid. 1989. *Architecture: A Place for Women:* Berkeley: University of California Press.

Boys, Jos. 1991. 'Is there a Feminist Analysis of Architecture?' *Women and Environment* 10.

Fowler, Pauline 1984. 'The Public and the Private in Architecture: A Feminist Critique', *Women's Studies International Forum* 7.

Hosken, Fran. 1987. 'Shelter Urbanisation and Change: A Feminist View', *Women's International Network News*.

Menon, A.G.K. 1997. 'The Contemporary Architecture of Delhi: A Critical History'. Discontinuous Threads: Conference on Memory, Freedom and Architecture in Contemporary India (Conference Papers).

Krishnaraj, Maithreyi. 1991. *Taking Stock, Women's Studies and Higher Education: A Symposium*. New Delhi: Centre for Women's Development Studies (28 September); published in Indian Association for Women's Studies. 1992. Newsletter nos 16 and 17, 1992: 1–4.

Varkey, Karula. 2000. Abstract of paper 'Universal and Contextual-Approaches to Design Education in Architecture', submitted for the conference, 'Academics and Architecture, Restructuring the Attitude'. Organized by the Mumbai-based periodical *Indian Architect and Builder*, November.

24

Gender and Science: The Fiction of Lila Majumdar

Ipshita Chanda

> *Science fiction is....about anything you want to happen to write about and using that particular convention and those particular forms. You can say certain things about whatever you want to talk about that you can't say using any other conventions.*

This definition was offered by Samuel Delaney (Delaney and Russ 1984) in a dialogue with science fiction writer and theorist Joanna Russ. I would like to invoke one of Russ' early publications, 'What Are We Fighting For' (1997). Here, she defines feminism as 'the study of patriarchal systems of unrecompensed labour and the political propaganda to maintain it' (Russ 1997: 253).

This essay begins with two premises: first, that science fiction follows particular forms and conventions of theme and structure that identify it. Second, that the genre is based upon the society that it seeks to leave behind through a progress facilitated by science; the fictional society upends the codes of the one the readers live in. For instance, in the science fantasies Lila Majumdar of (1908–2007) scientific temper and scientific practice are reimagined to address the

---

I am grateful to all the others in the group–this was one of the most 'active' email collaborations that I have participated on (facilitated by 'science'?), especially Gita and Sumi without whose constant support this would not have been done; and Moupia and Moromiya, without whose help it could not have been done.

needs of women and marginalized classes who, according to Russ, are victims of the patriarchal system of unrecompensed labour. Majumdar was born in a family at a time when the nurturing of a 'scientific temper' and the emancipation of women were taken as constitutive elements of an emerging modernity in Bengal. How does gender figure in her science fantasies? Can she be seen as shaping contemporary feminist discourse through her endeavour to write science fantasies from a gendered perspective? That is the question addressed in the chapter. I shall argue that Lila Majumdar establishes certain conventions for the genre of science fiction in Bengali which display a sensitivity to gender as well as a certain attitude to what we commonly understand–and misunderstand–as science. According to Attebery (2002: 1):

> Science fiction is a useful tool for investigating habits of thought including conceptions of gender. Gender in turn offers interesting glimpses into some of the unacknowledged messages that permeate science fiction.

Further, Attebery conceives of both gender and science fiction as a set of stable codes. Codes, he explains, have grammars (he might have also said, 'vocabularies') which facilitate translations into and from their repertoire of signs. They may be used to send or conceal messages. I would want to take issue with Attebery's contention that codes are stable: this is what he implies when he says that the existence of a grammar means that messages can be verified and analysed (ibid.: 2). I would suggest that rather than verification and analysis according to fixed standards, the codes simply mark out spaces within which fluid processes of identification, confirmation and resistance occur. I argue that gender inflects the discourse of science that is constructed by Majumdar's fiction and compromises the stability attributed by Attebery to the representational codes of science. The accepted codes which identify 'science' and the 'scientific temper' are refigured by Majumdar from a gendered perspective: but are notions of gender questioned or changed? To answer this question we may turn to the fictions.

Making the Knowledge One's Own

Batash Bari (BB, The Wind House)[1] and *Haoar Dandi* (The Wind Balance),[2] are two novellas for children by Lila Majumdar that she describes as 'science fantasies'. Indeed, Bengali science fiction is almost entirely for children. The question is, what is the difference between 'fiction' and 'fantasy'? Is the latter so-called because its audience is young, while that of the former is adult? This is a question that seems not to have been addressed in Bengali science fiction at all.

Susan Kray (2002) notes that science fiction is a particularly gender-conscious genre: its two main enterprises are 'having adventures' and 'solving problems' (ibid.: 37). This is strikingly applicable to the texts that I am discussing. In order to discern the way gender changes, reflected in the roles for women in Majumdar's science fiction, we may locate the characteristics of the genre in the work of the most popular writer of science fiction in Bengali till date, Satyajit Ray, who is Lila Majumdar's brother's son. Ray's science fiction hero is Professor Shanku (1983), a figure resembling Calculus (the absent-minded professor scientist in Hergé's adventure comic series, *Tintin*), whose exploits admit women only marginally, if at all. Shanku is a well-known scientist, with friends in the scientific fraternity the world over. He has adventures and solves problems with the help of science, using gadgets that he has himself invented. As a male scientist, this is natural to him. It is also natural that he is a single man without a family, and that in the entire series, there are no women to distract him or the (male?) reader.

Science for What and for Whom

In contrast to Ray, Majumdar's conception of the scientist even if he is male, is not dominantly masculine. In the case of Ray's Shanku, the scientist asserts his intelligence and proclaims himself as a genius. His experiments are important for themselves: whether they contribute anything directly to the greater good of mankind is not an immediate issue, and is rarely referred to in the story itself. In the story 'Nakurbabu and El Dorado'(NED) Shanku has been writing an article on all his work for a famous science journal *Cosmos*, where he notes that

> Whatever I have myself done, with the minimum expenditure, and the minimum of resources, the international scientific community has never been tardy in generously recognizing it. But there is another group of people who do not accept me as a scientist at all. In their opinion I am a magician or a shaman. I know many kinds of *mantratantra*, spells and enchantments, which will throw the dust into the eyes of scientists, and that is how I have earned my fame (NED: 66).

And Shanku also takes this opportunity to tell us what is unique about his work:

> Why have I not spread the word about my most important discoveries and inventions, like the annihilin pistol, or the Miracurol medicine, or the Omniscope or Microsonograph, or the machine called Remembrane which revives the memory–none of these can be built in a factory. These are all handmade, made by human hands, and those hands belong to just one man and none other. And he is Trilokesar Shanku (NED: 67).

So, the male scientist, almost a legend, battles against the odds, misunderstood, brilliant, astoundingly dexterous. He is a genius, and particularly, one not encumbered by any human relationships. As Russ (1995) argues, patriarchies 'imagine or picture themselves from a male point of view . . . other men and women in our culture conceive the culture from a single point of view–the male' (ibid.: 81). This seems to describe the world of Ray's science fiction, too. In Majumdar's novels, however, the location for the use of science, as well as for the adventure it provides, shifts from the laboratory and the dangerous jungles of darkest Africa. Majumdar's scientists and their work are located in Bangaon, a poor hill community where the basic necessities of life are hard to come by and science can either help or completely destroy the delicate balance of human being and nature. This location is far from the exotic or enclosed spaces inhabited by Shanku. Majumdar may call her work science fantasies but the protagonists are ordinary people threatened with

Gender and Science 257

ordinary problems, often caused by wrong applications of science. In the words of the Big Lama, who says in *Batash Bari:*

> The oil and coal under the earth will finish any day now. Twenty more years, at the most. Suppose you manage to get some more new quarries, then let's say forty years. . . then everything will dry up, finish off and become like a coconut that is too old to hold water (BB: 49).

And what will the scientist do, in these circumstances? Perhaps this is where the 'fantasy' enters, facilitating another view of science. In *Batash Bari*, Big Lama/Duley, the scientist, is still confident enough to say,

> As soon as the oil and coal are exhausted, I'll catch the sun's energy for you. If there is no paddy or rice, I'll drive away your hunger with pills. If there is a shortage of space, then I'll make a thousand wind-houses and hang them in the air (BB: 37).

Duley is emphatic about his engagement with nature and certain about using science in the service of his poor and hungry compatriots.

While Shanku's most useful invention seems to include the annihilin, which wipes out the enemy in a trice,[3] the aim of the scientists in both of Lila Majumdar's stories is of a different nature. This maybe because it is set in a different world. Kray (2002: 37) points out, that the work of the first women science fiction writers was marked by what psychiatrists call a 'feminine sensibility', and 'informed by a politics directly based on a traditionally patriarchal or masculinist agenda', because as a genre, science fiction does not deal with housework, child care, power status, work or fun. This may be true of science fiction written by women in European languages, but this is not true of Lila Majumdar. In her novels, food is Enid Blytonesque in its detail, perhaps because these are stories for children. Majumdar's concerns with the environment, the community and the domestic locations of the characters serve to locate science in a real rather than utopian or elite community. Science is seen as a force that can take man to this utopia by solving basic and common problems–not intellectual problems but practical ones which are generally not part of the code that science is defined by. For example, in '*Sejomamar Chandrayatra*' (Third Uncle's Moon Journey),[4] another Majumdar story, the protagonist Indra is a young boy who is being sent to the moon in a space ship. Before he embarks, the scientist gives him a pill that is supposed to be a nutrient. The young boy swallows the pill and then describes in detail the food he feels he has consumed (284): each taste is etched upon his tongue, so to speak, he can actually taste them all. Science is seen to be capable of replicating the taste of food: in this story, at least, the pill is not merely utilitarian. In *Batash Bari*, the children of the village go to the Big Lama when the mountain streams have dried up and the river is almost dead due to the lack of rains and hunger besieges the inhabitants of the mountain village. He too gives them a similar food pill, almost a staple of the science fiction convention. The Big Lama says that if they eat the pill they would not feel hungry for the next 24 hours. The pill will prevent them from having any food-related illness, too. But this does not please the children at all: they cannot forsake the sensuous quality of food, and complain that they want everything that accompanies eating, including the consequences:

> We want stomach pain, hunger, fishbone in throat . . . We'll share, smear it, pick out the bad bits, swallow, then only will it be fun, how long it is since we have smeared salt on anything (BB: 30).

This reminds the reader at once about the central problem that necessitates the use of science: hunger and poverty arising out of despoliation of the earth also caused by science. Indeed, even the solution is expressed in a metaphor related to food. The children see through a telescope that volcanoes and mines are not left to degenerate but 'people are filling up the empty stomach of the earth (HD:115). The Big Thakur in *Batash Bari* says that the invisible axis on which the earth tilts has lost its balance because humans have

> scraped out the earth's innards from its belly, the wind-balance is not getting the adequate weight to maintain its angle, and so it is

Gender and Science 259

gradually tending towards becoming a straight line, and so everything is turning topsy-turvy (HD: 102).

So,

The land slid down, trees fell, even the elephants left the blue mountains and began to move down. But there was no heat in summer, no rains during the monsoons. Last year even in winter, the winter wind did not blow, Why are the birds still not leaving this land to go south? (HD: 72)

Unlike Professor Shanku and his male scientist friends who are world authorities in their own fields, the children and the scientists in Lila Majumdar's work are 'ordinary '. They are made aware of the ways in which science as an instrument of greed has affected their lives and how they can combat the destruction. But this awareness takes on a new dimension against the backdrop of the usual mythification of science and the scientist. Shanku's frugal methods, and his insistence on living away from the public glare in Giridih, a small town in Jharkhand, do not come from the same outlook upon science that can cause the writer to locate a science fantasy in the almost inaccessible mountain village Bangaon, where the bounties of science have yet to reach the inhabitants. Shanku could have had fame and fortune, but he chose fame and a mystique, renouncing, like a mendicant hero, worldly pleasures for pure knowledge. The people in Lila Majumdar's novels have no choice. For them science is not an obsession. It has been used to changed their lives through depletion of natural resources and imposition of an inhuman order: they are struggling to reclaim their well-being using science. But the better life they seek is not one of isolated intellectual inquiry like Shanku's. For example, science is an obsession for the Big Lama too, but the difficulties of living in daily touch with poverty force him to look at its possibilities from a perspective that Shanku is insulated against. And this is from where Lila Majumdar's own vision stems too: as a woman in a society that is patricentric and masculinist, she imagines a utopia crafted by science that functions on what feminists would call equality and a woman of Majumdar's generation would call 'humanity'. The context and connotation of this statement may be clarified by Majumdar's autobiographical work, *Pakdandi* (The Footpath).[5]

> I had discovered something about myself. I cannot tolerate anyone's domination. If anyone orders me to do something, I turn recalcitrant, which means that I am not fit to take up any responsible position for an institution. I only look for a harmony of minds... If I think the other person has a rational argument then I am always prepared to change my position. But I have never been able to accept anything just because someone told me to. That is the root of my conflict with my father (226).

Domesticity, Ecology and Women

Russ (1995) points out that the plots that writers shape their fiction into are 'dramatic embodiments of what a culture believes to be true–what it would like to be true or what it is mortally afraid might be true' (ibid.: 81). Majumdar's science fantasies confront what the masculinist culture believes to be true with what she as a rational, humane woman–if not a feminist–would like to be true. Unlike Shanku who eschews family and companionship except for other male scientists, Majumdar's science fiction weaves in family and relationships as the background and the beneficiary of science. This makes for a difference in the stories themselves. For example, instead of portraying science through the heroic exploits of single, unattached men in quest of objective truth, both Majumdar's science-based novellas depict families and, especially, mother figures who are present even when they are absent. The children who come into contact with and learn to understand the wonders of science are inspired by the memory of absent mothers. So, when Pampa leaves her home because of ecological crisis, she takes a manasa plant to give her mother in heaven, certain that science will cause them to meet. Sondeo's mother is often remembered. The mother of Pema and Pemi has been left behind by the villagers of the high mountains when they retreat to the foothills, scared by the ecological disasters that have destroyed their way of life. While these mother figures ensure that food is served, eaten, described and enjoyed they are also protectors and provide a sense of security to the community in times of trouble. For instance, when the temple is abandoned by the frightened villagers because their priest has told them that the demons will come in, drive God out and take over the temple, Pema's mother, despite her lame leg manages to go up to it, ring the bell and worship there. It is the heroism of a woman fighting on the side of rationality against the superstitions of the village men. She stands alone in the temple as a human being staunch in her own faith rather than cowed by fear of the unknown. These women collectively indicate the background of values and questions against which science is practised and its fruits socialized. Majumdar makes the point that science is socially located intellectual activity, and human beings are not robots–they have active agency in the control and direction of scientific practices, and may use them to dominate or alleviate the sufferings of common people.

Joanna Russ (1972: 88) says that the work of female science fiction writers of the first generation, beginning in the 1940s, 'contain more active and lively female characters than do stories by men,' but they are generally space-age variants of 'ladies magazine fiction' in which a 'sweet, gentle, intuitive little heroine solves an interstellar crisis by mending her slip or doing something equally domestic after her big, heroic husband has failed'. Turning to Majumdar's stories, however, we see that the domestic and the scientific are not seen as discordant. This may be because the imbalance of the ecological system threatens the very lives and livelihoods of the villagers. There are no narratives of women's empowerment or heroism: rather, when science is summoned to correct this imbalance, women, who are 'naturally' seen as guardians of the domestic sphere, actively participate to redirect science towards preserving the values of care and nurture which they embody. In the stories, women are involved like all other members of the community in preserving lives and livelihoods, and neither their gender nor their effective functioning in roles that the 'real' world denies them, is remarked upon. Whosoever is able to help in redirecting science for the benefit of the poor villagers, be it man, woman, girl or boy, plays a part in the narrative. But the process by which women reached the unusual roles they play in the narratives, or even an acknowledgement of their unusual nature, does not appear to concern Majumdar. She offers no explanation for this, and therefore raises the question whether she is a feminist writer.

Russ (1995: 81) argues that novels depend upon what action can be imagined as being performed by the protagonist–what can the central character in a book do? Examining English and American fiction, she comes to the conclusion that 'of all the actions that people can do in this fiction, very few can be done by women'. It is true that the women are not at the centre of Majumdar's stories; rather, when the adventurers arrive at their destination, the women are already there. True too that what they have done in order to get there and their achievements are taken as a matter of course. By not remarking upon the presence of the women who have already arrived at the destination that the young characters in the novellas want to reach with the help of science, Majumdar's science fantasies appear to reverse the convention of celebrating the achievement of women protagonists in explicitly feminist fiction. For example, Diya drives the spaceship that brought the children to the Wind House. They have come here to correct the tilt of the earth's axis and restore the wind balance, so that the ecological system stabilizes. That she is a woman seems incidental as the narrative does not place added emphasis on her gender, but makes her present to us as an active character. Majumdar seems to present a world in which the definition of women's roles have changed: as indeed they must if science is to achieve its true aim of benefitting human beings. It appears from the novellas we are discussing that Majumdar naturalizes for her young readers the changed roles of women in this new society where science is being harnessed to alleviate human suffering rather than amass wealth and power.

The story of the empowerment of women which expanded their roles from that of nurture and care in the domestic sphere to that of scientific competence in the public sphere does not seem to her to be worth telling. Does she then want her young readers to take the new roles of women as an obvious consequence of the new idea and practice of science itself? Her narratives make the fictional society resulting from a different scientific practice not only one in which women 'naturally' perform roles they did not in the earlier society, but also one in which the perceived 'feminine' values of nurture and care form the base for the practice of science. The results of science first appear as life-threatening to the people of Bangaon; but the adventure the narrative takes the characters on is one of discovering that science can also be used in ways that are apparently mundane. These uses of science in this changed society serve to mitigate the ecological disasters that follow if science is allowed to bracket out the values seen as feminine: so just as it is natural, in the new imagined society for women to drive spaceships, so it is natural for science to find safe ways to conserve water. For example, when they wash up after eating in the Wind House, the water turns to steam through solar energy and returns to the bottle (BB: 65). The latter is shown to be an achievement of science which can be noticed only if we are concerned about water resources and their availability, which the rampant misuse of science has placed under threat. So the values underlying the practice of science have changed, though the participation of women is not explicitly cited by the author as the cause for this change.

In order to present this change, Majumdar has changed the conventional codes of science fiction. Imagining as Delaney requires, a world where what is not possible in this one can happen, she places science within a social frame. Hers is an ordinary world where ordinary people have to bear the brunt of science misused, a world where science can furnish the answer, if human beings choose to use it for common good rather than for personal profit. If humans choose to abuse the earth, 'suck it dry until the seed is exposed to the elements' (HD:105), then they are to blame. They alone can change the situation: but even though they are scientists, they are not omnipotent, nor omniscient. In *Haoar Dandi* the Big Thakur says:

> I am not responsible for telling you the solution. I can only tell you why this is happening and what else will happen if we continue to do what we are doing. There is a huge sum at the root of everything, and I have only learnt to understand that (HD:101).

The very attitude to scientific knowledge has been addressed in very simple terms in both these stories. One might shrug it off by saying that these are children's stories, and the author herself has called them fantasies. But perhaps because of this, the power play of patents, intellectual property rights, transmission of the scientific discovery and the invention are quite haphazard. Through this, the ownership of scientific capital is refigured to benefit the 'unrecompensed labour' of the common people who have and use this knowledge for common good. In *Batash Bari*, the scientist Duley allows his inventions to be credited to his employer, Nepen, who has made the money for his experiments and inventions available. The problems regarding ownership of knowledge are directly addressed by the author. Since the Batasi company has funded the making of the robot, should it not be owned by the company? How can Duley tamper with it though he is its actual maker? This introduces the relation between capital and knowledge as a product, albeit in a concretized form, that of robot 49. As Duley says to Nepen:

> This [robot called 49] is in no way the property of the Batasi Company. It has been built with my life and mind and the money from your salary (BB: 55).

To the suggestion that the Batasi company had provided materials for building the robot, and so should own it, Duley retorts:

> They provided materials, as if! I ask, did they give the sunlight? The brains? (BB: 33)

This makes short work of the legalistic interpretation of the fact that everything has its price. And science fiction can become the vehicle for problematizing the claims of transparency, functionality and objectivity that science makes for its own discourse. In Majumdar's work, this problematization occurs through reimagining the aims and idea of science itself. What role does gender play in this endeavour?

The perspective of gender is used by Lila Majumdar to demystify both science and its operations, while not demeaning or idealizing them in any way. We are faced with the fact that scientists are far from perfect: employed as a cook, Duley dishes up inedible food, even though he is a first-rate scientist. Perhaps domesticity, held to be mindless repetition and the work of women, defeats him. But scientific knowledge also presents a limited, one-dimensional world. Take, for example, the Big Thakur's ignorance about weaver birds' nests. He does not know that they have the entrance to the side away from the storm winds (HD:111). When told this, he retorts, 'I know only sums: I don't know weaver birds' (ibid.). If the feminist reader asks why the Big Thakur has to be a man, one may reply with this detail, which reveals the blindness of specialization to quotidian wonder. As a powerful man who can control science, he is unaware of simple natural things known to children. Thus Majumdar takes the responsibility of humanising the scientist, not by changing his gender, but by putting him in situations in which his human limitations and his human qualities are exposed. Quite unlike the uncompromising and equally unremarked, taken for granted masculinity of the milieu of scientific practice that we find in the Shanku stories, Majumdar's protagonists, men and women, are social beings whose gender roles have changed in the new society, resulting from the practice of science for human benefit. For example, in Majumdar's '*Shejomamar Chandrayatra*' the scientist Kunal Mittir's son Manohar makes a spaceship that takes him to a farm instead of the moon. This is the kind of mistake male scientists in Majumdar's work seem to be capable of. Yet Kunal Mittir, Manohar's rational scientist father proclaims to the neighbours that the boy had made a very good spaceship, only evil people were spreading rumours about him (SC: 289). So, scientists have feelings of irritation, outrage and pride in their offspring's idiosyncrasies, like ordinary human beings.

In a story entitled 'Night of Fire' by Helen Reid Chase, published in 1952 in the short-lived science fiction journal, *The Avalonian*, on the apocalyptic 'night of fire', the aliens make good their vow of saving the best of humanity, quietly whisking away scientists, engineers, and housewives alike. Chase celebrates the possibility that women in the home—much like men in the laboratory and on the assembly line—might contribute to a new techno-cultural world order. But, as we have seen, Lila Majumdar's characters inhabit an ordinary world and the change they wish for is fundamentally related to their very survival. The brave new world that they envision is one which will give them equal access to the basic necessities of life. The earlier environment may not have provided equality, but there were adequate resources for all. However, the misuse of science as an instrument of greed has changed the situation and led to a depletion of resources and now equality of distribution is the central issue. It is at this juncture that 'modernity' enters the plot: the ideas of equality and justice are the central tenets of this 'enlightened' civilization, the dawn of modernity in Bengal.

The effect of this upon women will be clear from Majumdar's autobiography, *Pakdandi*. Besides the accounts of innumerable female lives of varied qualities, prejudices and generosities, the author's own analysis of the position of women in her society is worth noting. Writing in the early 1970s, she says

> Fifty years ago . . . if not all, at least two thirds of the women who would move freely in public, were Brahmo girls; moving freely meant that they would go to college without a male protector, not one in their wildest dreams imagined going to the cinema. Among the Hindu boys these girls met, in 90 per cent of the Hindu households, and in 80 per cent of the families the men came from, the women of the family did not speak to men 'other' than their husbands. It was not impossible for these men to imagine that it was impossible for the women who studied alongside men to remain 'good' (209).

The domestic space inhabited by Brahmo Samaj women of Majumdar's age and class in the early twentieth century is complementary to the public space which women entered as they were educated and came forth to do the 'work' of the nation. The movement of these women across these two conjoined spaces was not easy and often much pain was caused by contradictory responses from the society within which their individual and collective lives were lived. These pressures were reflected in Majumdar's relations with her father.

> When my father broke caste and married my mother [who was a Hindu], my grandmother and elder paternal uncle agreed [to the marriage], though they must certainly have been unhappy. But my father's happiness was the most important to them. But almost thirty years after all this, when I married my Hindu husband through registration under the Gour Act, I did not get my father's consent, his blessings or his forgiveness (118–19).

It is this liberal paternalism that the author rebelled against: but it was also something that she records as respecting. As pointed out earlier, this is the period when modernity was shaped in response to the western ideas that the colonizers brought: but the local traversed the field implicating all dominant social discourses, among them the discourse of gender and science.

The operation of gender in Brahmo society was markedly different from its operation in Hindu circles. New social relations were forged due to the functioning and the beliefs of the Brahmo Samaj. Rabindranath Tagore (1966), member of a family at the centre of the Brahmo Samaj in cultural and commercial power, describes this change thus:

> We were ostracized because of our heterodox opinions about religion and therefore we enjoyed the freedom of the outcaste. We had to build our own world with our own thoughts and energy of mind. We had to build it from the foundation.

It is this 'own' world of the elite Brahmo Samaj in constant interaction with upper-caste Hindu neighbours and English colonizers that forms the structure of feeling in Majumdar's fiction. She has fashioned her ideas of science and its operation in society against this backdrop. The connection of science with society genders Majumdar's fictional world, while Ray disregards society and gender completely. Majumdar's introduction of gender precedes the discourse of science that became dominant through Ray's work. Her meagre output of science fantasies shows us a view of science and a world in which science operates very different from the masculinist world of science in the Shanku novels and stories. The world of science still struggles with feminist critiques of the objectivity and heroism of male scientists, and their structuring of science as an exclusively male domain. That Majumdar is trying to demolish this world in her science fiction is revealed by many generic inversions she makes, some of which have been pointed out above.

Another characteristic of the fictional world she creates for the practice and the understanding of science, is the presence and participation of children. Though both Ray's and Majumdar's works are marketed as 'written for children', and children appear in their science fictions, their actual narrative function is vastly different if we compare the two sets of texts. Both Majumdar's stories include children, girls and boys, who are interlocutors and participants in the adventure of science. Lisa Yaszek (2008: 3) argues,

> In direct contrast to the feminist fiction that authors…[like Russ]. were beginning to produce, women writing for the postwar SF community rarely seemed to take the next logical step and show how new sciences and technologies might produce new sex and gender relations as well.

As we have seen, Majumdar, as a pioneering science fiction writer in Bengali, redraws the roles played by men and women in scientific endeavour by placing them in the context of society and the family, rather than in isolation from both, as Ray does with Professor Shanku. Through the introduction of a human community as participating in, being oppressed by and benefiting from science, Majumdar's science fiction for children presents a different world made possible by a located and humane practice of science.

Majumdar's childhood in the Shillong hills makes her sensitive to environmental degradation: and the use of science in both ecological destruction and preservation is a survival issue rather than an intellectual pursuit for its own sake. The scientific advances that wreak havoc on people's lives and lead to despair and misery are evident in the hills being mined and deforested to fuel the industrialization of modern India. State-regulated industrialization was based on a model of development that now stands discredited: but Majumdar writing more than thirty years ago, questions this very ideal. She makes the human suffering that science has the potential to cause, the reason for a beneficial, scientific quest. Majumdar locates her stories in the Himalayas, in an ordinary milieu where care and nurture are ordinary virtues, amongst families and friends. In this milieu, science can be of service, but till date, it has only plundered these regions. The basic necessities are denied to the hill people because of the despoliation of the earth: machines often become threatening because they take away jobs from people, who then starve. This is what happened to the workers in the Batasi company where Duley's master Nepen worked. 'Nepen had a robot who worked equal to 15 people. Factory workers were being laid off, so everyone protested. Nepen loosened a few screws, and the result was that the robot did whatever it liked.' But should machines then be banned? In reply, Padam, one of the children in *Batasher Bari* says: 'Men will work on the land. There is so much to do there' (BB: 29).

Does this suggest the beginnings of an alternative discourse of science from a feminist perspective? We have attempted to craft a reading inflected by the author's experience of gender in society at a particular time and place. These inform her conception of science, its practice and philosophy. Science is used in the fictional world created by Majumdar to produce a structure of feeling which is not obtainable in the real world. The real world has misused science, but the fictional world proposes an alternative, beneficial use. This is possible in the world of science fiction, where many operations not possible in the present world can occur due to some ingenuity and skill produced by science itself. But such a world imagined by Majumdar is not the same as the world constructed by popular science fiction writers like Ray. We are attempting to understand this difference through the interlinked discourses of gender, science, family and society. Let us now turn to the 'scientific' temper shaped in this discursive habitus. What are its markers and how does the woman writer inflect it with her gendered perspective, if at all she does so?

Consider, for example the ability to ask questions. This could perhaps be called the truly scientific temper from a particular location within society: but it is precisely what someone else from another location within the same society at the same time may well ascribe to philosophy. Hence, the criteria of separation of one from the other is fluid both synchronically as well as diachronically, and must inform our reading of social discourses translated into the codes of literature, or what Attebery would call genre. To take an example, the construction of relationships in each of the texts is upon human lines, whatever the alien looks like. These relationships, which form the dynamics of the narrative and make it flow at a particular pace in a particular direction, seem to demand the understanding of certain codes of inter-personal behaviour. Lila Majumdar considers, perhaps naturally, the relationship between the robot and its inventor in this perspective. As Duley/Big Lama says about his robots, 49 and Dhingipada,

> The maker's entire intelligence is in the belly of the robot, from complicated mathematical formulae to the sudden start that one experiences when the kite calls shrilly from the top of the date-palm (BB: 49).

Whether this move on the part of the author will qualify as 'feminist' or not is an open question. But as we have seen from the critical interventions of feminist science fiction theorists and practitioners like Russ (1984), Lisa Yaszek (2008), Helen Merrick (2009), Ursula LeGuin (1994), even when women themselves write science fiction and consciously try to change the parameters of representation, they fall into what Russ describes as a suburban trap (1971: 81): the known world is their coordinate, and often, the female scientist replicates gender roles prevalent in patriarchal society. How shall we deal with this slippage, both as readers and critics, and perhaps also as feminists?

Considering gender and science as mutually articulated codes Atteberry (2002) was able to show how genre and gender interacted within narratives rather than constructing literature, science and gender as separate spaces. Lila Majumdar prompts us to take this perspective by according both reason and imagination equal and interacting space in refiguring the codes representing science in her work. When the children in *Batash Bari* begin to start talking about food even after eating a nutrient pill, the Big Lama is irritated: how is this possible? They can't be hungry already? The pill's effects are supposed to last for a longer time. The children have a different explanation for it however. They tell him:

The hunger in the stomach is cured by pills, but how to cure the hunger of the heart I don't know (BB: 26).

Is this a scientific attitude, or is it the recuperation of science for another agenda? Has the focus shifted to emotional need, something that is commonly seen at the opposite pole from the popularly perceived rational objectivity of science itself? If women refuse to embrace scientific and social change, Chase (cited by Yaszek 2008: 2) warns, at best they will be left behind and at worst driven insane by the demands of a world that is rapidly evolving past them. And this brings to light another possible opposition. It seems that this opposition should be between Reason and God. Usually, this opposition is identified as Reason *versus* Faith. But it is the deification of reason in a particular guise that has closed off the possibilities of admitting positions that lay claim to rationality. Science's claim that objectivity and empirical verifiability define knowledge has gone some way in aiding this deification. A contrast is provided by the conception of science offered by the characters in Majumdar's science fiction. For example, Sondeo explains this in his own way: 'Sums are another name for God. It is that which is at the base of everything: weaver birds too.' (HD). Are robots capable of enjoying whatever they do for our benefit? Aren't they supposed not to have any feelings? By attributing human feelings to them, is Lila Majumdar attempting to redraw the conventions that define 'human' and 'machine', or is she doing away with those conventions altogether? And this is where the 'scientific' temper of her science fiction may be located. The idea of 'science' that informs her work may be gleaned from such instances.

Let us place this 'science' in the context of Russ' contention that the feminist narrative may be a social documentation of the fact that capital is built upon unrecompensed labour. We may only add that this is possible when one assigns an exchange value to labour itself, such that both physical and mental labour are capable of being measured, and exchanged. The difference between human and machine 'labour', and the relative determination of the worth of each is indeed a part of Lila Majumdar's narrative, as an answer to an obvious question, a reason that may be affixed to a form of just differentiation. This is a question about the relative worth of the work of machines and men. The Big Lama/Duley explains it thus:

> The work of the machine is never wrong. All its calculations are always correct. Not a single line is even the tiniest bit crooked, it is only human hands that can make mistakes. That is why the work of human hands commands such a high price (BB: 23).

This prompts us to think that the way a society evaluates labour and ascribes material exchange value to it is reflected in the organizational ideology of that society. Or we may see science as either facilitating or obstructing the solution to an obvious problem which causes so much strife in the world: the equitable distribution of wealth, which Majumdar addresses directly as arising from the lack of control over the fruits of science. Satyajit Ray's work provides an exact example of what happens when a prevailing generic code is used to write science fiction in Bengali. Ray's science fiction, completely unlike his (paternal) aunt's, is chockfull of all that feminists are immediately up in arms about, and for very valid reasons. His is a world in which women, domesticity, care, nurture are absent, and where one of Shanku's most important inventions is the annihilin, which wipes out the enemy completely within seconds. We have attempted to use Ray's work which is the most popular example of the genre in the language as a contrast to Majumdar's. Majumdar had started by redefining the genre in most of its essentials. But we have not yet considered the actual sense of wonder and the power of human faculties that she identifies as the epitome of the scientific temper.

Feminist Utopia or a Different Science

Lila Majumdar seems to weave a narrative that shimmers like the 'dragonfly spaceship' (HD: 102) that she herself shows us in *Haoar Dandi*. It is 'made of wind and air, and it sparkles'. When describing machines, and certainly while describing energy, the pictures that flash upon the page dissolve the senses into what seems to be a sensuous experience. Perhaps the process is triggered off by what

Gender and Science 273

Joanna Russ, another writer of science fiction in a completely different chronotope, describes thus:

> What finally happens is that what you thought was a metaphor isn't one. It is literal reality. What we know about the universe comes to us through that light....the light of the stars is all we know, and most of our knowledge is light itself because we are not nocturnal, we are diurnal animals. Something like 80 per cent of our imagination comes to us through sight. So what you have....is not a metaphor but a literal reality and the whole experience becomes cleansed and refreshed and beautiful (1984: 35).

Majumdar offers us many such reports of experience: 'Blue means nothing but distance. First blue, then grey, then it disappears' (HD: 75). Is it the strength of this language that the 'truth-value' and scientific provenance of each of the images translates into what we call poetry? For this is the same language in which the dynamism of the world and the futile desire to control it through ownership are simultaneously evoked.

The padre has taught the children in the first book a song taught in the church and sung there:

> The forest god moves
> Leaves fall from the trees
> Young leaves are made to bud
> The silk-worm pupa is filled
> Thick sticky sap trickles down (BB, 34).

Attebery (2002: 15) contends that science fiction 'challenges accepted notions of nature and culture incorporating signs from within science and technology in such a way as to evoke sensation of strangeness: not novelty, but a reordering of categories. This view aptly describes Lila Majumdar's science fiction. But is this reordering of categories related to altering the codes of science 'fiction' itself? Attebery in fact prompts us to consider this question by describing science fiction as a 'metascience, examining different ways of knowing'. He calls it an essentially realist mode, because it 'convincingly constructs faked histories'. The identification of 'reality' and 'the fake' become ambiguous if we apply them as categories to understand Lila Majumdar's writing. Which are the foundational categories for understanding her work? The position of women in the real and the science-fictional worlds, or the ecological disaster and poverty caused by human misuse of science in one of them, which the writer of science fiction claims may be ameliorated by the proper use of science in the latter? We have argued that these are related through experience of social being since gender is a part of lived social experience. Majumdar uses science to create an utopia: she thinks of her novellas as science fantasies, because clearly, they are not stories of the 'real' world. And gender equality is just another element of this world of fantasy. The conventional gender roles and the masculinist ideas of science are refigured by making women 'natural' actors in a new–and fantastic–society where the feminine values of care and nurture are foundational to scientific practice. The writer sees this scientific practice as socially located and beneficial to the poor and the ordinary–quite different from the intense intellectual fervour which excludes all human society except that of other male scientists, so prominent in the mainstream science fiction of Satyajit Ray. Thus the metascience of fiction is used by Majumdar to critique the existing discourse of science and suggest an alternative discourse, both made possible through the perspective of gender.

The value of such an exercise, as we have undertaken in this essay, it has to be acknowledged, is the mapping of a road. As Attebery says

> And because where we are going depends partly on where we are willing and able to imagine ourselves going, science fiction can offer insights into the limits of the imaginable and the ways those limits are changing (Ibid.: 15).

Science fiction for all its ambiguities extends the imaginable. Lila Majumdar's science fiction directs the imagination to a different view and thus a different use of science through the use of gender as a perspective from which to understand and change the existing world. From the gendered perspective, she imagines a world where science ameliorates the hardships of living. In her fiction, science produces and is produced by a worldview in which care and nurture as well as family and friendships are not marginal but integral to the practice and aims of science. This extension of the imaginable world crafts a new discourse for science and for literature in the language. Majumdar writes about the conservation of solar energy with a food metaphor that is delicious as it is literally warm and sweet, like mellow sunlight. Since the earth will soon turn into a coconut that is too old to contain water, the Big Lama wants to conserve light using the principle that is used to make a mango sweet called *amshatta*. Here the amshatta will be made out of light: the colours of the light will be diffracted and conserved like the thick juice of the mango is squeezed and spread, layer upon layer, on a surface exposed to the sun to dry and congeal into a chewy, sweet amshatta. In every layer of this amshatta made of colour, the energy of sunlight will be conserved. People will live in wind houses suspended in air, or up in the hills where they are closest to the sunlight, so that they can use these luminous amshatta for their energy needs.

Is this a feminist utopia, or one constructed by objective, rational science? This chapter has attempted to show that Lila Majumdar's Bengali science fiction for children identifies the scientific temper as the mentality which recognizes that science can imagine and make possible a just and equal world by introducing the gendered perspective into the genre in Bengali literature. However, this task once begun has not been continued–as we have noted, even Satyajit Ray with whom she worked closely, did not follow her lead. He preferred to use the accepted codes of a heroic male scientist in an apparently genderless world–and he remains till today the most popular science fiction writer in Bengali. Majumdar's early step on the road to a gender sensitive discourse of science, remains till date a journey that has not begun.

Notes

1 Lila Majumdar, *Batash Bari* (BB) (1984), originally published serially between April 1973 and January 1974 in *Sandesh*, a children's magazine

that Majumdar and Satyajit Ray edited until the latter died in 1992. The edition used here is from *Lila Majumdar Rachanabali*, vol. 5 (Kolkata: Asia Publishing Co., 1984). All references to this story occur henceforth in the text as BB, with page numbers following. All translations from this and all other Bengali texts used in this essay are mine.

2 *Haoar Dandi* (HD) (1984), originally serialised in *Sandesh* between April 1982 and January 1983 the edition referred to here is collected in Lila Majumdar *Rachanabali*, vol. 5 (Kolkata: Asia Publishing Co., 1984).

3 See, for instance, '*Shonkur Congo Abhijan*' in Ray, 1983: 46, where a gorilla is killed using this pistol, or '*Mahakasher Doot*' also in Ray 1983: 18.

4 Lila Majumdar. 1955, '*Sejomamar Chandrayatra*', in *Lila Majumdar Rachanabali*, vol. 3 (Kolkata: Asia Publishing Co. 1955): 279–86. All page references are to this edition.

5 Lila Majumdar. [1986] 2008, *Pakdandi* (Kolkata: Ananda Publishers); all references are from the 2008 edition. All translations are mine.

References

Attebery, Brian. 2002. *Decoding Gender in Science Fiction*. London: Routledge.

Chase, Helen Reid. 1952. 'Night of Fire', *The Avalonian*, 1; cited by Lisa Yazsek. 2008, *Galactic Suburbia*. Columbus, Oh: Ohio State University Press.

Delaney, Samuel, and Joanna Russ. 1984. 'A Dialogue: Samuel Delaney and Joanna Russ on Science Fiction', *Callaloo* 22, Fiction: A Special issue (Autumn): 27–35.

Kray, Susan. 2002. 'What Women Don't Say'. In *Science Fiction: Canonization, Marginalization and the Academy*, edited by G. Westfahl and G. E. Slusser, Westport CT: Grenwood Press: 37–50.

Le Guin, Ursula. 1994. 'Coming Back From the Silence', interview with Jonathan White, http://www.swarthmore.edu/Humanities/pschmid1/engl5H/leguin.interv.html Accessed 13 July 2011.

Majumdar, Lila. 2008. *Pakdandi* (Kolkata: Ananda Publishers).

———. 1984. *Batash Bari*. *Lila Majumdar Rachanabali*, vol. 5. Kolkata: Asia Publishing Co.

Majumdar, Lila. 1984. *Haoar Dandi*. In *Lila Majumdar Rachanabali*, vol. 5. Kolkata: Asia Publishing Co.

———. 1955. '*Sejomamar Chandrayatra*'. In *Lila Majumdar Rachanabali*, vol. 3. Kolkata: Asia Publishing Co.

Merrick, Helen. J. 2009. *The Secret Feminist Cabal: A Cultural History of Science Fiction Feminisms*. Seattle, WA: Aqueduct Press.

Ray, Satyajit. 1983. '*Nakurbabuo El Dorado*'. In *Shanku Ekai Eksho*. Kolkata: Ananda Publishers: 66–98.

Russ, Joanna. 1997. *What Are We Fighting for? Sex, Race, Class and the Future of Feminism*. Gordonsville, Va: St Martins Press.

———.1995. 'What Can a Heroine Do? Or Why Women Can't Write'. In *To Write Like a Woman: Essays in Feminism and Science Fiction*. Bloomington, IN: Indiana University Press 79–93.

———.1972. 'The Image of Women in Science Fiction'. In *Images of Women in Fiction: Feminist Perspectives*, edited by Susan Koppleman Cornillion. Bowling Green, OH: Bowling Green University Popular Press: 79–94.

Tagore, Rabindranath. 1966, *Talks in China Autobiographical*. In *English Writings of Rabindranath Tagore: Plays, Stories Essays*, the section *Autobiographical*, edited by S. K. Das, vol. 2. New Delhi: Sahitya Akademi: 583.

Yazsek, Lisa. 2008. *Galactic Suburbia*. Columbus, OH: Ohio State University Press.

25

Gender, Science and Technology Education in India

Anitha Kurup

'The paths leading to a science career are as diverse as the people traveling them.'
—Jill Bargonetti (2008)

The Human Development Report 2004 emphasizes the need to respect diversity and build a more inclusive society. By adopting policies that reflect cultural identities and encourage diversity, a country will make greater strides towards development. Diversity in the Indian context will have to address not only the gender dimension, but also categories of caste, class and rural/urban differences. I explore the concept of diversity using gender as a category; the arguments developed may be extended to the other categories and be nuanced to deal with the interplay of caste and class with gender.

The participation of women in formal education at all levels reflects inequality. Higher education, particularly science and engineering education, is not very different. Against this background, the focus of this chapter is on unpacking gender imbalances in science and technology (S&T) education with specific reference to India. The attempt is to contextualize the history of women's engagement in the field of S&T that has been characterized by exclusion and inequality within the larger framework of diversity because of its potential in building a more inclusive society as stated by the *Human Development Report 2004* (UNDP 2004). The report calls attention to diversity and cultural identities, two aspects which could be cornerstones of the development of any nation, and is the only *Human Development Report* to do so.

India's populations belong to different religions, castes, languages and classes with diverse experiences and cultural specificities that have the potential to contribute to its growth and development. A multicultural approach, therefore, is not only desirable but viable and necessary. Diversifying the pool of scientific and technological talent in India has to take into cognisance the dilemmas of equality and quality on the one hand and keeping pace with the advancement in the field on the other, while responding to micro-level problems. The inclusion of women to extend the potential of diversity, thereby providing India with a competitive edge in the global S&T scene, has to be seriously considered. In this effort the mechanisms for democratic participation of women may have to be redefined, so women are not only end-users, but also producers of scientific knowledge. This will in a way, create a more conducive environment, where an equitable role of women within S&T institutions can be reclaimed. What are the major challenges in this process and how could these be addressed? At one level, it may be important to examine women's role both in terms of numbers/presence as well as the nature/process of their participation. What do the existing research studies both at the national and international level indicate?

Research Studies on Women in S&T

Research Studies on women in S&T are isolated and spread out in terms of geographical coverage. They also represent different points of time. Perhaps the most comprehensive study was the 'Science Career for Indian Women: An Examination of Indian Women's Access to and Retention in Science', which probably for the first time, attempted to gather and analyse data at the national level (http://insaindia.org/executive.htm). This study has revealed useful information particularly with regard to the proportion of women at senior and management levels as well as in the academies. In addition, the report highlighted gender-insensitive organizational practices, workplace gender-related discrimination, nepotism and even sexual harassment. However, the nature of discrimination was more indicative than illustrative and nuanced. The low representation of women at senior and decision-making levels has figured as an area of concern (Poonachand Gopal 2004; Bal 2004). Despite the growing numbers of women in biology, prevalent patriarchal practices hamper the career prospects of these women despite these women having a relative advantage of education and class when compared to women in the unorganized sectors (Bal 2004). A few others who carried out case studies with reference to specific universities and institutions referred to tacit factors–organizational and professional rigidities, reflective of social attitudes, hindering the progress of women in a science career (Gupta and Sharma 2002; Kumar 2001; Subramanyam 1998). A joint study (Kurup, Maithreyi, Kantharaju and Godbole (2010) of the Indian Academy of Science (IAS) and the National Institute of Advanced Studies (NIAS), Bangalore, explored the reasons why women after completing a PhD in science drop out from scientific research. Absence of job opportunities, institutional factors and the opaque selection process were among the several reasons. The proverbial glass ceiling that cuts short a woman's career in science is not the family but the systemic biases in S&T institutions. Particularly in the Indian context, the prevalent structures and practices of these institutions that have historically been a male domain pose a serious challenge to women's balancing career and home. Despite this, several women scientists have found their own approach to balance career and family in interesting ways. These experiences have not been studied and analysed so far.

Research studies on women in S&T in the West (especially countries like the USA, UK, Germany, Sweden, Australia and many others) have had a relative advantage because national-level data at all levels of education up to the PhD and beyond are available. This data is sex-disaggregated across various broad disciplines and sometimes at the sub-discipline level even within science and engineering. The data are updated and subjected to analysis, which in most cases informs planning and policy making at the national level. The data within disciplines and across years are useful to understand trends in participation. The revelation of disaggregated numbers has been instrumental in several initiatives undertaken by these countries to address the problem of gender imbalances in S&T. Despite consistent efforts over more than two decades, however, the persisting gender gap is a great concern in the USA and the European Union among other countries (European Commission 2011; Hill, Corbett and St. Rose 2010; National Academy of Engineering 2005; and Seppo 2010). Compared to India, the interface of the women's movement with the scientific community has had a longer history in the West. Although India has had a very vibrant women's movement, its interface with the scientific community or for that matter engagement with issues related to women in science has been relatively recent. Feminist discourses among women scientists are yet to be established, and this has resulted in either the women scientists attempting to study women's issues within a limited framework or sociologists and women studies scholars independently conducting studies with inadequate understanding of the complexity of the scientific institutions. Their impact on the science policy makers of the country has been limited, as in the field of science the validity of the results is dependent on large representative data sets.

Recognizing the problem of the gender gap in science, India can critically review the efforts taken by the western countries and choose those measures that are relevant to our context. Or India can make a concerted effort through collaborative surveys and research studies that will bring together the scientists and social scientists to understand better how gender interfaces with S&T institutions. Engaging the scientists in a constant dialogue even to agree to disagree is a useful tool if one has to impact the science policy in the country. Keeping communication channels open through constant negotiations; recognizing the differences and idiosyncrasies of disciplinary backgrounds; and developing mutual respect are important methods to engage in meaningful research across seemingly disparate disciplines. Policy recommendations that come out of these collaborative efforts are likely to have a greater impact on women in science.

Among the major challenges that India faces in this direction is that the data are unavailable, dispersed or often outdated even after sixty years of independence. Furthermore, the lack of comprehensive statistics and comparable gender-sensitive indicators are serious obstacles in studying the situation of women in science in India. It is surprising that while there are serious efforts to bridge the gender gap in education at the lower levels of formal education, where gender as a category of analysis is receiving some attention, the same is not true about higher education, especially in the context of science and technology. Going by the estimates of the World Bank (2002), that by 2015 there will be 97 million students enrolled in higher education and that half of these will be in the 'developing' world, there is no attempt by India to provide updated gender-sensitive and gender-disaggregated data that will help us squarely address the problem of the gender gap in higher education. The need for the gender-wise data by disciplines at all levels of education including the PhD level is crucial to make any meaningful comparative analysis.

In addition, the qualitative experiences of women who enter the domain of higher education remain largely un-researched and un-theorized. This is also true of disciplines in S&T. Examining data of practising women scientists and engineers not only limit the analysis but does not give us a reference point in relation to their male counterparts. The process of engendering, therefore, needs to go beyond numbers and focus on other factors. There is also a lack of serious engagement between three distinct bodies of literature on gender, development and education, particularly in relation to higher education. Indeed, overcoming this may hold the key to provide dynamism and directions to the engendering process in science and engineering.

Finally, the enormous expansion of ICT (Information Communication Technologies) throughout the 1990s began to change both the world economy and the place of higher education institutions in this new economy. Apprehensions have been raised on borderless universities, offshore, franchised, satellite and on-line learning and the expanding global reach of higher education. These debates are not critically informed by gender and this is a matter of concern.

Re-emphasizing diversity and establishing its link to building more equitable S&T institutions may be useful at this point. There is a close link between diversity, representation and engendering science and technology. That women comprise 50 per cent of the human population necessitates equal representation in all spheres, S&T included. The exclusion of women is a matter of concern not only from the point of view of fairness, nor even of inadequate numbers of trained personnel, but because of not harnessing the experience of half of humankind, resulting in making S&T that much poorer. The under-representation of women in science and engineering has received relatively less attention in India, despite being recognized as a serious problem.

Diversity Key to Scientific Progress in India

The challenges of addressing diversity in India stem from the high levels of inequality in our society. Higher education, particularly S&T institutions, reflects this inequality, which therefore becomes institutionalized. Thus, one is confronted with inequalities both among different types of institutions and within each institution. The uneven distribution of resources amongst centres of higher education has more than one dimension. Compared to the humanities and social sciences, the natural sciences have and continue to receive a larger proportion of financial resources. This unequal distribution of resources derives from post-independence policies that rested on the belief that advancement in scientific development would contribute to the development of the nation as a whole.

Even within the natural sciences, however, the distribution of resources is uneven not only between institutions but also across disciplines. Disciplines like atomic physics, nuclear energy, space science and more recently nano sciences and biotechnology, continue to attract greater funds compared to foundational disciplines like physics, chemistry, botany, and zoology. This is also true of disciplines that have a social component like the agricultural sciences, conservation studies, or environmental sciences. While one may like to argue that privileging some of these disciplines also reflects masculinization, it is not clear whether the privileged discipline by itself qualifies for being categorized as masculine and hence attracts more resources; or whether the ability to attract resources results in drawing an increased number of men as compared to women. Whatever be the rationale, it is a fact that the few women who remain in scientific research are relegated to disciplines that attract relatively less resources. Cutting-edge research in laboratory-based experimental science is cost-intensive. Restricted funds for research in certain biological sciences (excluding perhaps sub disciplines like nano biology or biotechnology) that have a relatively higher number of women, directly impacts the research output. Equal distribution of resources among the higher education institutions is an important issue for diversity.

Some centres for higher education are among the best both in terms of attracting resources as well as the quality of research produced. There are many others with a comparatively larger student outreach (like the state universities) that have neither funds and infrastructure nor human resources to carry out research of international quality (Yashpal 2010). There are many more women in the state universities than in the more exclusive research institutions. The disparity between a large number of universities on the one hand and islands of excellence on the other is so large that these exclusive institutions attract the best students, faculty and enormous resources, which put together produce quality research. But at what cost? Depriving universities of research funding cannot be justified by any means.

Universities that have in the past been nerve centres of research in S&T have now been reduced to teaching universities with extremely limited resources for research. The growth of parallel autonomous research institutions supported by the Council for Scientific and Industrial Research (CSIR), Department of Science and Technology (DST) and Department of Biotechnology (DBT) have been able to attract large funds, which has had a negative impact on universities, particularly in relation to research activities. Unless the country addresses this imbalance in the distribution of resources, the country will fall short of qualified human power that is required for its rapid growth and development. The paucity of quality scientists and engineers is already apparent. Qualified S&T personnel cannot be trained and produced by the miniscule number of institutions of excellence. Building on the existing infrastructure of universities and augmenting them with substantial resources for research will benefit the nation in the long run.

In more recent times, diversity has assumed a different angle where the point of reference is human resources. The *Human Development Report 2004* emphasizes the need to respect diversity and build a more inclusive society. By adopting policies that reflect cultural identities and encourage diversity, we will make greater strides towards development. A multicultural approach is desirable, viable and necessary. In India, institutions of higher education whether the mass-based universities or institutions of excellence have their own histories and draw students from different socio-economic and cultural backgrounds. The degree of diversity is much less in elite institutions when compared to universities. It has been argued that increasing the diversity of students by gender, class and caste will have a positive impact on quality in the long run as it provides rich opportunities for collaboration and mutual sharing of experience. The impact of diversity on equality in higher education in India needs to be assessed. In the Indian context, diversity will have to include not only the gender dimension, but also categories of caste, class and the rural/urban differences. However, this chapter will explore the concept of diversity using gender as a category to illustrate the point. The arguments developed can be extended to the other categories and will require to be nuanced to represent the interplay of caste and class with gender.

Evidences of Diversity

Diversity has attracted increased attention in the West, especially in the USA and Sweden. Diversity of the student population by gender and ethnicity, and amongst the faculty has also been promoted (George 2003; Daniels and Warren 2003). Efforts are underway in a planned and structured way to meet the growing demands for professionals in S&T during the coming decades. Multifaceted strategies have been adopted that range from conducting research to understand the past trends better, examining more closely the relationship between diversity and performance of universities across the country. Data in the USA between 1983 and 1996, the period when engineering recruitment strategies were adopted to promote participation of women, resulted in a visible increase in the number of women enrolled in undergraduate engineering courses. Programmes adopted by Purdue University, for example, resulted in a rise of enrolment of women from under 1 per cent in 1969 to 5 short of 1000 five years later (Meiners 2012). Comparable efforts with different strategies to increase the number of women in science and engineering have yielded positive results elsewhere in the West (Ivie and Ray 2005).

The argument for diversity has more than one dimension. Promoters of diversity very often view it within the framework of fairness or social justice. They argue that in a world that seeks to be just, it is only natural that there is proportional representation of the different categories be it gender, race, ethnicity and the like. Others argue in terms of the projected requirement of a higher number of trained professionals in S&T that it is only natural to draw the additional numbers from the excluded groups. This dimension of the argument is more pressing in the case of India and other Asian and African countries that witness a large outflow of trained scientists and engineers to the developed nations in search of greener pastures and better working conditions (Kurup 2006).

An interesting third dimension of diversity proposed by Wm A. Wulf (1998) is that diversity promotes creativity. Although S&T has always projected itself to be logical and abstract in its design and content, it is the creative angle of these disciplines that makes them more fascinating. Creativity in science has by far been the greatest contributor to humankind. Wulf (1993) has also argued that creativity in engineering is the reflection of one's whole life experiences and hence excluding the life experiences of half of the world's population makes science poorer. As a result, 'we pay an opportunity cost—a cost in products not built, in designs not considered, in constraints not understood, and in processes not invented' (Wulf 1998).

In his discussion on diversity in engineering, Wulf brought to the fore the importance of encouraging multi-disciplinarity, not only in defining a problem but also in seeking challenging multifaceted solutions. This requires scientists and engineers to move beyond the narrow confines of disciplinary boundaries and draw from other seemingly unrelated disciplines both at the conceptual and methodological levels. The recent breakthrough in disciplines like nano-sciences that bring physics and chemistry together in more fundamental ways along with other related disciplines of instrumentation and electronics are striking examples in this direction. The Human Genome project and the cognitive science initiative in India demonstrate the potential of drawing from different disciplines. The Department of Science and Technology's cognitive science initiative, for example, has brought together, cognitive scientists, neuroscientists, behavioural scientists, psychologists, mathematicians and social scientists to interrogate questions in relation to cognition in humans and animals.

The increasing diversity at the individual level demonstrated by scientists and engineers in the recent years may have significantly contributed to the emergence of new disciplines or rather areas of enquiry that fall between them. This has, in turn, enriched the body of knowledge and made significant strides in the realm of applications. At another level, it has also set into action a new academic culture that has created a relatively more level playing field for women and other underrepresented groups who hitherto have not been a part of the traditional academic science and engineering community. The presence of women within the scientific community in these relatively new disciplines may have widened diversity if not by numbers but by the very difference in the nature of experiences that form a part of the scientific endeavour. This, unfortunately, may not be true in the case of traditional disciplines in science, given the dominant paradigm of thinking during that period, when women may not have had an opportunity to make contributions. The introduction of a diversity of 'experiences', in relatively newer fields of specialization needs to be analysed more carefully before drawing any definite conclusions. The motivation, engagement and confidence of women scientists may have increased just by drawing upon the paradigm of relational learning predominantly used by women (Brooks 1986). The interface of an axiomatic learning tradition (that defines rigour) with a relational learning tradition (through argument and proof) has resulted in solutions of problems that may have otherwise been limited by the use of either learning paradigm. It can he hypothesised that the greater role of relational learning is encouraging more women to specialise in the emerging fields of S&T. (For a more detailed discussion see Williams 2000.) Of the small percentage of the population who are hierarchical rather than relational learners, a majority are men and hence this framework not only excludes women but more importantly a large percentage of men. If diversity encourages relational learners, this could hold the key to competitive S&T in today's world and be worth examining.

Not only would more students from diverse backgrounds be attracted to these critical branches of knowledge, but also their presence and contribution may open new vistas in the knowledge base leading to overall progress in scientific development. Here the emphasis is not to undermine the importance of hierarchical learning in pursuing science but opening the disciplines to other equally important modes of learning, enhancing quality rather than hampering it. Myths created around underplaying the potential of diversity also need to be revisited. It is often believed, and sometimes even systematically argued, that broadening access and allowing entry of students from diverse backgrounds impacts quality negatively. On the other hand, it is argued that diversity has a positive impact on quality. Both these propositions are anecdotal and have to be supported by substantial research. There is need for large scale systematic research before propagating either proposition. Evidence from the West shows that diversity does not impact quality of education negatively, rather the change is positive (Murray 2005; Clifford and Royce 2008).

It may be a fruitful exercise to document if there is a variation in the composition of student population amongst institutions in India, and if so analyse the underlying reasons. The data demonstrate that there are strong educational reasons for universities to recruit and admit a diverse student population. The advantage of student diversity is further enhanced by faculty who recognize and use diversity as an educational tool, by including content related to diversity, reorganising content and adopting methods that create an inclusive and supportive classroom climate. The educational outcomes are comparatively more superior (Antonio 2003).

Faculty Diversity—An Important Dimension

Diversity in the Indian subcontinent has received less attention. The historical advantage to formal education that the privileged sections of society had in relation to caste, class and gender brought little conflict as the student population in an earlier period was homogenized to a great extent. Notwithstanding the positive discrimination policy adopted by the newly formed Indian government that opened doors for the underprivileged sections, the overall composition remained more or less the same in the case of science and engineering, which occupied a relatively higher status. The recent expansion due to the demand for higher education coupled with encouragement for private enterprise in higher and professional education has resulted in small changes in the diversity of students at least in certain S&T disciplines. The faculty composition reflects less of this diversity and shows variation by institutions and also across disciplines (Eubanks and Weaver 1999).

One of the most prevalent barriers to progress toward a diverse faculty is recognising and getting beyond myths. Myths around the issue of diversity represent presuppositions that may be held by institutions and some recruitment committee members at the conscious or unconscious level. Myths contribute to maintaining the status quo. To illustrate, *'Diversity is only for minorities'* is a myth that that creates resistances in a group of practicing scientists. However, all can benefit from exposure to diverse perspectives (Turner 2002).

Under-Representation of Women

Improving scientific, technological and vocational education is most important if girls are to improve their lives and if the problems of poverty, particularly in developing countries, are to be tackled. Even in countries where girls are well represented in science courses at the undergraduate level, there is a distinct pyramidal effect, with the number of women dwindling significantly at the doctoral and post-doctoral levels. Achieving gender equity in science requires that girls have access to basic science education from an early age and that they receive mentoring at the secondary school level that will make them aware of possible careers in science—particularly in new fields such as computer science, biotechnology, and environmental science—and will enable them to make the educational choices necessary to open the doors to those opportunities. Graduate women can play an important role in this process serving as role models and developing mentoring programmes that will counteract stereotypes, make mathematics and sciences more attractive and encourage girls to enter careers in science, engineering and technology.

The question of engendering science and engineering has often been formulated within the narrow scope of participation per se. Data on enrolment at different levels are analysed to drive home the point that participation of girls at all levels of S&T education is progressively increasing. However, disaggregated data in terms of sub-disciplines, across different levels and, more importantly, over the years, are not recorded and even when this is done, the data are not subjected to further analysis. The point I would like to raise in relation to understanding the participation of women in S&T has three levels.

One is that enrolments of students are not independent of other players who constitute an important component, namely, the teachers/faculty. This is true at all levels. Hence, it may be important to look at enrolments along with recruitment of faculty. Lack of data support in general poses a challenge in locating underlying factors as well as the recognition that gender disaggregated data are the key to derive insights. The real answer to this national problem may lie not in limiting it to a gender question, but a question that examines the nature of the student and teacher population and the extent of its homogenisation. In this analysis, there is an urgent need to study the shift in the student and faculty composition across different disciplines over a period of time. Tracking the career paths of students and faculty, especially at the higher levels, beyond the doctoral and post-doctoral, is crucial. The shifts in employment of faculty, both men and women, need to studied in depth to at least attempt to make any meaningful interpretation of the trends.

Two, researches carried out in this area have used dominant paradigms which have very often shifted the focus of attention onto the girls' abilities and factors like parental pressure, individual choices, long working hours and the inability to balance career and family life. The over-emphasis on societal and individual factors to explain away the poor presence of women in science and engineering leads one to believe that all other factors including institutional are neutral and provide a level playing field. This is not to underplay the effect of societal factors but to recognize the institutional culture that may have a bearing on the real access to fields in S&T. In other words, there is a need to review the paradigm by questioning the very propositions of the direct relationship between individual ability and societal factors on the participation of girls in S&T. While faculty may constitute an important component, other factors not so obvious like performance measures, work culture, and so on, may also be contributing factors.

Three, the inadequate analysis of the experience of women scientists and engineers, mostly resulting from adopting traditional frameworks, has failed to throw light on intrinsic issues that may provide an explanation for the gender imbalance in S&T. Research studies have more often than not treated women scientists as a homogeneous group, whether from the point of view of socio-economic backgrounds or for that matter their period in the profession. The spectrum of experiences of women scientists must include those who have experienced little or no barriers in career advancement in academic institutions, those who have experienced barriers but are not conscious of or sometimes unwilling to admit this, and those who have identified barriers but have not been able to overcome these in their academic lives. Redefining new categories may pave the way to resolving or at better understanding the issue of gender balance in the larger interest of development of S&T in the country.

There is a need to shift the question of women's participation in S&T from ways in which women can be motivated or encouraged to redefine the problem within the realm of gender relations. This will mean understanding male participation in S&T is as important if not more so, to get to the crux of the issue. It is imperative that research studies move beyond the organizational culture of these institutions to addressing the lack of intertextuality between three distinct bodies of literature, namely gender, development and education.

Problems related to data include their lack in the form required, delay in making the data public and inconsistencies with regard to sources as well as definitions, posing serious problems for comparisons. The purpose of this chapter is not so much to dwell on the limitation of data availability but to draw attention to the causes of the problem of women's under-representation in science and engineering. It is true that available data in the current form limits the analysis and conclusions drawn, but more important is that there is an urgent need to try and understand the processes that lead to the gender imbalance as indicated by the data. Thus, data in the current form are indicative of a larger problem and not the problem per se.

Engendering Science: The Way Forward

The body of literature on gender and science points to the lack of sustained qualitative data in virtually all the studies analysed so far–the complexities of alienating organizational cultures are not always recorded. Gupta and Sharma (2003), in their study of gender inequality in the Indian Institutes of Science and Technology have listed among other elements the tacit biases and stereotypes that operate in the application of rules. The gendering of research opportunities requires one to take note of this. Access to students, projects, funds are affected due to bureaucratic hurdles. Apart from these the study also points out that being part of informal networks is integral to success in academics in science and engineering. Kurup, Maithreyi, Kantharaju and Godbole (2010) in a recent study, have drawn attention to organizational factors as being responsible for women dropping out of science. Only between two and five per cent of the women scientists (of a sample of 568 women scientists surveyed) report organizational factors as responsible for their quitting their earlier jobs. A nuanced interpretation of this result, however, becomes crucial against the background of the sample of personal interviews conducted. The low reporting of organizational factors could be due to various reasons. Most scientists believe that in science, its practice and organization are always fair; indeed, the onus of carrying this image is perceived to rest on women scientists.

At another level, given the closed organization of scientists and the in-built hierarchy amongst this group, getting scientists to report on organizational factors may be difficult due to fear of the possibility of tracing the response to the scientist. Balancing family and a research career is a challenge that most women in academic careers face. Combined with other social and family factors such as caste, class, geographical location, and so on, these factors play out in various permutations and combinations and result in several women opting out of research careers.

While changes at the family and societal levels are slow and painstaking, a way forward for S&T institutions in this scenario would be to effect changes at the organizational level. Scientific institutions can offer flexibility in timings, as well as options to work from home through access to internet and online journals and libraries at home; making more provisions for housing on campus or arranging for transportation; rescheduling meetings and other such events during the morning hours; making provisions for good quality childcare facilities at work, and so on, as the first few steps. In contrast to research institutions and universities, several corporations have already recognized the valuable contributions of women and offer several of these provisions including flexibility in timings, work-from-home options, childcare facilities or transport. However, it needs to be emphasized that scientific organizations have not been studied in the Indian context. It may also be that the organizational culture is taken as given and has not by and large been a subject of enquiry. Attempts to understand this black box have not received adequate attention. Drawing an analogy, in slightly different but related contexts, this has also been the case in India with schools as an institution up to the eighties. There is a need for a greater volume of work raising questions beyond traditional frameworks that may allow one to have a peek into S&T organizations.

Our study also revealed that the National Science Policy in India has to take note of the diversity among women scientists, namely, women engaged in science research (WIR), women not engaged in science research (WNR) and women who are currently not employed or working (WNW). Significant differences were found in the provisions considered useful to retain women in science research among the above three sub-groups of women scientists. While all three sub-groups perceived flexibility in timings to be useful, more WIR have perceived provisions of accommodation and transportation to be useful. More WNR have perceived human resource policies to be useful, while more WNW have perceived provisions such as job security, age limit relaxations, congenial working atmosphere, etc. to be useful. The differences in perception perhaps reflect the particular experiences of the individual sub-groups and the nature of difficulties they face in balancing careers with their other responsibilities. The data importantly reveals the need for policies to refrain from developing blanket-provisions that do not meet the needs of all women scientists, but perhaps address the needs of those already in science. Further, the differences seen with respect to the different age-cohorts importantly indicate the need to reorganise decision making and 'standing committees' at the national and state levels to promote the participation of women in science and also younger age groups so their views and experiences are also represented.

As already mentioned, there is a lack of intertextuality between gender, development and education. The literature on gender and that on development treat education as one component; within development literature education receives relatively less attention than the more obvious issues like poverty, livelihood and health. The literature on women in higher education is limited not only due to inherent data problems but also the lack of understanding of gender analysis often only restricted to enrolments. There are very few qualitative studies in the public domain built around women's experiences and engagements with higher education outside the West. There is some literature on equity and higher education in the UK (e.g. Bagilhole 2002; David 2003; Eggins 1997; Howie and Tauchert 2002; Leonard 2001; Morley (1999, 2003a). Efforts to understand the interface of gender in institutions of higher education particularly S&T are still sparse. The interface of gender with education in the Indian context is of recent origin. This is true of most developing nations. We need to also engage with the knowledge construction and practice of science. Addressing the inter-relationship of gender, development and education may be an important step towards enhancing inclusion, equality and diversity in S&T in India.

References

Anitha, B.K. 2006. (See also Kurup, Anitha) 'India's Competitiveness and Preparedness in Science and Technology in the Coming Decades–A Meeting Report.' *Current Science* 90, 3: 281–283.

Anitha, B.K. and K. Kasturirangan. 2007. (see also Kurup, Anitha) 'Engendering Science and Technology Education in India.' International Conference on Women's Impact on Science and Technology in the New Millennium, 2005, Bangalore. Ed. V. Krishnan, TWASROCASA, JNCASR, Bangalore, India.

Antonio, Anthony Lising. 2003. 'Diverse Student Bodies, Diverse Faculties'. *Academe* 89, 6: 14–17.

Bagilhole, Brabara. 2002. 'Challenging Equal Opportunities: Changing and Adapting Male Hegemony in Academia.' *British Journal of Sociology of Education* 23: 19–33.

Bal, Vineeta. 2004. 'Women Scientists in India: No Where Near the Glass Ceiling.' *Economic and Political Weekly* 9, 32: 3647–52.

Bargonetti, Jill. 2000. 'Beating the Odds: Remarkable Women in Science', Science AAAS in partnership with L'Oreal Foundation Enterprise, AAAS Science Business Office, 'The paths leading to a science career are as diverse as the people traveling them'. http://www.amit-es.org/assets/files/publicaciones/beating%20the%20odds.pdf Pg: 7 Downloaded 17 September 2012.

Campion, Patricia and Wesley Shrum. 2004. 'Gender and Science in Development: Women Scientists in Ghana, Kenya, and India', *Science Technology and Human Values* 29, 4: 459–85.

David, Miriam E. 2003. *Personal and Political: Feminisms, Sociology and Family Lives*. Trentham: Stoke on Trent.

Clifford, Derek and Maureen Royce. 2008. 'Equality, Diversity, Ethics and Management', *Social Work Education: The International Journal* 27, 1: 3–18.

Eggins, Heather, ed., 1997. *Women as Leaders and Managers in Higher Education*. Buckingham: Open University/SRHE Press.

Etzkowitz, Henry, and Namrata Gupta. 2006. 'Women in Science: A Fair Shake?' *Minerva* 44: 185–99.
Eubanks, Segun C. and Reg Weaver. 1999. 'Excellence Through Diversity: Connecting the Teacher Quality and the Teacher Diversity Agendas.' *Journal of Negro Education* 68: 451–59.
European Commission. 2011. 'Addressing the Gender Gap in Science and Technology.' *Innovation Union Competitiveness Report 2011*. (http://ec.europa.eu/research/innovation-union/pdf/competitiveness-report/2011/chapters/part_ii_chapter_3.pdf at 1.15 Downloaded 8 July 2011.
Gupta, Namrata, 2007. 'Indian Women in Doctoral Education in Science and Engineering: A Study of Informal Milieu at the Reputed Indian Institutes of Technology.' *Science Technology and Human Values* 32: 507–33.
Gupta, Namrata, and Arun K. Sharma. 2002. 'Women Academic Scientists in India.' *Social Studies of Science* 325/6: 901–95.
Hill, Catherine, Christianne Corbett, Andresse St. Rose. 2010. Why So Few? Women In Science, Technology, Engineering, and Mathematics. Washington, D.C.: AAUW.http://www.aauw.org/files/2013/02/Why-So-Few-Women-in-Science-Technology-Engineering-and-Mathematics.pdf Downloaded 28 September 2015.
Howie, Gillian, and Ashley Tauchert, eds., 2002. *Gender, Teaching, and Research in Higher Education: Challenges for the 21st Century*. Aldershot: Ashgate.
Indian National Science Academy. 2004. *Science Careers for Indian Women: An Examination of Indian Women's Access to and Retention in Scientific Careers*. http://www.insaindia.org/Scienceservice/science.htm. Downloaded 25 August 2012.
Inter-Academy Council. 2006. *Women for Science: An Advisory Report*. IAC publications. http://www.interacademycouncil.net/Object.File/Master/11/051/Part%201%20WFS.pdf Downloaded 25 August 2011.
Ivie, Rachel, and Kim Nies Ray. 2005. *Women in Physics and Astronomy 2005*, R-430.02. AIP Report. February.http://jurp.org/sites/default/files/statistics/women/women-pa-05.pdf. Downloaded 27 September 2015.
Kumar, Neelam. 2001. 'Gender and Stratification in Science: An Empirical Study in Indian Science', *Indian Journal of Gender Studies* 1: 51–67.

Kurup, Anitha, and R. Maithreyi. 2011. (See also Anitha, B.K.) 'Beyond Family and Societal Attitudes to Retain Women in Science', *Current Science* 100, 1: 43–48.

Kurup, Anitha, R. Maithreyi, Kantharaju B., and Rohini Godbole. 2010. *Trained Scientific Women Power: How Much Are We Losing and Why?* Bangalore: Indian Academy of Sciences Publications.

Leonard, Diana. 2001. *A Women's Guide to Doctoral Studies*. Buckingham: Open University Press.

Meiners, William. 2012. STEMing the Engineering Brain Drain. https://engineering.purdue.edu/WIEP/Spotlights/steming-the-engineering-brain-drain. 25 August 2012.

MIT. 1999. A Study on the Status of Women Faculty in Science at MIT. The MIT Faculty Newsletter, 11(4).http://digitalcommons.usu.edu/cgi/viewcontent.cgi?article=1016&context=advance 7 September 2012.

Morley, Louise. 2003a. *Quality and Power in Higher Education*. Buckingham: Open University Press.

———. 2003b. 'Gender Equality in Commonwealth Higher Education, The European Network on Women and Higher Education Conference, Brussels, June 2003.

———. 1999. *Organising Feminisms: The Micropolitics of the Academy*. London: Macmillan.

Murray, S. 2005. 'Workplace Diversity: Leveraging the Power of Difference for Competitive Advantage', *Research Quarterly*. Society for Human Resource Management.

NAE (National Academy of Engineering). 2005. 'Diversity in Engineering: Managing the Workforce of the Future'. By the Committee on Diversity in the Engineering Workforce.

National Academy of Engineering. Washington, DC: National Academies Press. www.nap.edu/books/0309084296/html/ 17 September 2012.

Poonacha, V., and Gopal, M. 2004. INSA Report, Research Centre for Women's Studies, SNDT Women's University, Mumbai, 2004.

Seppo, Roivas, ed., 2010. *Meta Analysis of Gender and Science Research*. Country Group Report, Nordic Countries—7th Framework Programme. DG Research, European Union. http://www.genderandscience.org/doc/CGR_Nordic.pdf 17 September 2012.

Subramanian, Jayashree. 2007. 'Perceiving and Producing Merit: Gender and Doing Science in India,' *Indian Journal of Gender Studies* 14(2): 259–284.

Subrahmanyan, Lalita. 1998. *Women Scientists in the Third World: The Indian Experience*. New Delhi: SAGE.

Turner, Caroline, Sottelo Viernes and Samuel L. Meyers. 2000. *Faculty of Color in Academe: Bittersweet Success*. Allyn & Bacon Publications, Minnesota.

UNDP. 2004. *Human Development Report*. New York: UNDP.

Williams, F.M. 2000. Access and Merit: A Debate on Encouraging Women in Science & Engineering. In New Frontiers—New Traditions, a National Conference on the Advancement of Women in Engineering, Science, and Technology. 2000. St. John's: NSERC/Petro-Canada Chair for Women in Science and Engineering, Memorial University of Newfoundland.

Wulf, Wm A. 1998. 'Diversity in Engineering'. In *The Bridge*. Winter 28(4): Revised version of talk given at the National Academy of Engineering meeting: 4 October. http://www.nae.edu/Publications/Bridge/CompetitiveMaterialsandSolutions/DiversityinEngineering.aspx Downloaded 25 August 2012.

Yashpal. 2010. Report of the Committee to Advise on Renovation and Rejuvenation of Higher Education, Ministry of Human Resources Development, Government of India. http://www.academics-india.com/Yashpal-committee- report.pdf Downloaded 25 August 2012.

26

An Inclusive Science and Technology Education Curriculum at School Level

Sugra Chunawala and Chitra Natarajan

The world today is increasingly dependent on science and technology (S&T) and the question of gender and S&T becomes even more pertinent. We not only require people to contribute to the growth of S&T, we require that they understand the impact of S&T on our lives and use its products effectively. That worldwide there is a trend of fewer women participating in the growth of S&T is worrisome, as it prevents inclusive perspectives. Education is viewed as a panacea for the imbalances and a harbinger of positive change from the existing status quo. Despite the emphasis on educational reforms, in India the societal picture has not changed. Indeed, research reveals that across the world education is gendered (AAUW 1992; Sadker, Sadker and Donald 1989).

Science as a subject has been a part of school curriculum for over a century and the gendered aspects of science education have been studied over decades. The under-representation of women in science and technology education has been a concern. A multiplicity of factors from our social environment, like family, media and peer interactions promote gender stereotypical roles to children; the school itself is an institution where gender identities are established. Some of the factors within the school, which have been identified as resulting in gender differences are: language of textbooks/curricular material, classroom interactions and the image of science and technology presented to students (ibid.).

In comparison to science, technology has not been universally accepted as part of school curricula. At present, technology is largely introduced at higher levels and is a subject at the school level only in some countries. Even so, it has always been gendered and perhaps more so than science. Layton (1993: 33) suggests that '"gendering" of experience is nowhere more obvious than in technology'. Technology tends to be perceived as complicated and not suitable for women, and engineering is seen as a masculine profession (Chunawala and Ladage 1998; Rosser 1992; Harris 1997). In this sense technology parallels science. Cockburn (1983, 1985) and Wajcman (1991) suggest that the relationship between technology and gender is two-way, where each is the source and consequence of the other. Thus the kind of technology we have is a reflection of the gender relations in our society, which in turn evolve to reflect our technologies. Wajcman (1991) writes that the technological content of women's activities is not appreciated or is undervalued. For instance, women's traditional involvement in technologies, such as horticulture, cooking, sewing and child-care, has been accorded low status.

In higher education, few women enter technological fields–a reflection perhaps of both the nature of these fields and gender relations in society. In India, women form a small part (about 22 per cent at graduate level) of the technology/engineering community (INSA 2004); of those who clear the engineering examination, over 30 per cent remain unemployed (Parikh and Sukhatme 2004).

This chapter attempts to advance the thesis that introducing technology education at school levels is essential to counter the gendering of technology. The worldwide trend of less female enrolment in science and technology needs to be challenged and the appropriate place to begin to do so, including the existing gendered practices of science and technology, is not at the higher education levels but at school.

We start by elaborating on the human contexts that are responsible for gendering S&T. Next we present the gendered perceptions of S&T and report on some studies from India and across the world. Then we focus on pedagogical practices, textbooks, classrooms, and research into the gendering of these. The following two sections summarize the historical origins and present status of S&T education in India and address concerns of equity and excellence. We then outline a possible scenario, where technology education in India is made inclusive. We suggest that education needs to focus on creativity through collaborative work among learners, and communication between and within groups of learners. The value of collaboration and inclusiveness has been upheld by feminist perspectives and is shared with other marginalized groups. Finally, we envisage future directions.

Science and Technology: Embedded within the Human Context

Both science and technology are human endeavours with a long history. They are products of social relations and forces, and hence gendered. According to Weber and Custer (2005) there is a skewed involvement of women at all levels of science and technology and a lack of visibility of women's contribution in S&T. The contribution of women to technology is 'hidden from history' (Wajcman 1991): the prototype inventor is male. The perception that what women do is not in any sense technological persists despite women's involvement in survival technology since the dawn of history.

The National Focus Group on gender issues in education (NFG-G&E 2006: 64) has emphasized that

> Various socio-cultural factors keep women from entering fields that are overtly called technology. Craft unions have played an active role in resisting the entry of women into trades, thereby relegating women to unskilled jobs and identifying skilled work with men.

Besides, technology selection is shaped by social arrangements reflecting the power structures in society. Political scientist Langdon Winner (1986) argues that technologies are not politically neutral. Their design may consciously or unconsciously include or exclude social groups.

Unlike technology, which is universal and a part of all cultures and groups, science is a specialized form of knowledge, acknowledged to be possessed by a few. It has a formal language and method, and implicit rules for membership to groups that possess and advance that knowledge. Philosophical debates about what constitutes science have a long history, and some consensus has been reached on the nature of science. According to Keller the modern view of science as masculine can be traced to the seventeenth century with the establishment of the British Royal Society. The society's purpose was 'to establish a Masculine philosophy . . . whereby the mind of Man may be ennobled with the knowledge of solid truths' (Keller 1985, cited in Haggerty 1995: 7). That science is viewed as masculine today is undeniable. Studies aimed at discovering attitudes to science reveal that the physical sciences are considered more masculine than the biological sciences (see also Chapter 17). The latter, considered more of a helping science and more people oriented, is dubbed 'soft'. It is no wonder then that girls are found more often in biological streams (Jones and Wheatley 1988).

The term 'technology' is a slippery one, and has context-dependent meanings, which have evolved over time. According to MacKenzie and Wajcman (1999), technology has three layers of meaning. The first refers to physical objects, like the cart, bicycle, refrigerator and computer. The second includes the human use of the physical objects, like cooking, grinding, driving and computing. The third is the 'know-how', or the information required to use, repair, design and produce the physical objects. These layers of meanings indicate the complex nature of technology. Philosophers are still grappling with the content of technology. But with respect to even the first layer, artefacts, there is gendering of various kinds and at various levels (Faulkner 2001). Designers, who envisage the structure and function of artefacts (such as watches for women or men) play an important role in the gendering. But this gendering can be altered by users who interpret and communicate about the artefacts (Oudshoorn, Saetnam and Lie 2002).

Technology and language, including words, signs and symbols share several characteristics. They are both tools of culture and located in the actions of persons and groups. They evolve when persons and groups participate in and negotiate their way through new situations. Technology, for instance, evolves in problem situations and these situations are located in a historical context. Language and technology enable as well as constrain thought, intellectual processes and action. Using a language in a certain manner serves to define a person or group's identity. Using tools (and technology) in a certain manner implies adoption of a cultural belief system about how the tool is to be used. Language production, meaning making, discourse, tool use and tool making are all best understood as a dynamic interplay between individuals and society at various levels of interaction.

Gendered Perceptions of S&T across the World and in India

Images change with time and place. Historically, images of men and women and their gender roles have changed and have accordingly been justified on different grounds. These justifications have ranged from the irrational to what some might term 'psuedo-scientific'. Images of science and its relations to gender have also changed historically (Schiebinger 1989). Feminist theory examines in detail how gender ideology permeates the social construction of knowledge and the use of 'genderized' language as it continues even today in science.

How do students view science and scientists? And is this perception important to science educators in any way? Chambers (1983) stated that in the eighteenth and nineteenth centuries, there were varied visual and verbal images of scientists, which are rarely seen now. Despite being stereotypical, the images showed diversity: representing villainy, clumsiness, madness, foolishness and eccentricity. However, as science has transformed its organizational structure, improved its general social status and has established its social authority, a new professional image has emerged in the media, which is a 'cleaned up' and standardized one.

The work of Margaret Mead and Rhoda Metraux (1957) with high school students in the United States indicated that students view science as natural science, and the scientist as 'a man who wears a white coat and works in a laboratory, is elderly or middle aged and wears glasses'. The methodology involved required students to write essays in response to questions. Joan Solomon and her colleagues (1994), by using questionnaires and interviews with students in Britain aged between 11 and 14 years, obtained varied images of science. They included a cartoon image of science/scientists; scientists as vivisectionists; all-knowing; as technologists; teachers and pupils as scientists; and scientists as entrepreneurs. The use of drawings to explore students' images of science has a long history. The 'Draw a Scientist Test' (DAST) has the advantage that it does not depend on verbal facility and hence can also be used with very young children. This test was devised and used by Chambers on around 5000 children between 5 and 11 years old, during 1966–77. The test has also been used widely by researchers in different countries. Newton and Newton (1992) found that children acquired stereotypical images of scientists, as early as six years of age. These studies indicate that science is largely perceived as a masculine domain by students despite efforts to present science as gender-neutral.

According to Sjoberg and Imsen (1988) an image of science is a cumulative result of various school and out-of school influences. A study was undertaken in Mumbai with eighth standard students (about 13 years old) as part of an international collaborative effort coordinated by Professor Svein Sjoberg (Norway), Dr. Jayashree Mehta (India) and Jane Mulemwa (Uganda). This study with students from Mumbai used a multi-task approach to gain insight into students' ideas about science and scientists from various perspectives (Chunawala and Ladage 1998). The tasks required students to use different skills ranging from merely checking, to writing essay type answers and drawing. One of the most consistent findings was an overly positive image of science and scientists. Students' positive image of science and scientists may be viewed as a favourable manifestation in a developing country like India, which since its independence has emphasized science and technology for development. A very positive image of science and scientists has also been reported by earlier studies of students in the West (Mead and Metraux 1957). Some recent studies, however, show that students associate destructive and harmful aspects with science, and hence they do not hold a glorified picture of science (Solomon, Duveen and Scott 1994).

The Mumbai study (Chunawala and Ladage 1998) reported that students' image of science and scientists was a stereotyped one: a young, intelligent, hard working male, a solitary person engaged in laboratory work, most often chemistry. Biologists were viewed as neater, more caring, social and kinder than physicists, who were viewed as more intelligent, imaginative, hard-working, interesting and democratic. This stereotype of science and scientists matches earlier reports (Sjoberg and Imsen 1988).

It is interesting that Indian students did not often draw scientists as old, with a beard, wearing glasses and a lab-coat. Their image is closer to reality than the western stereotype of a scientist. The laboratories drawn were also neat and tidy in appearance, rarely showing signs of danger and accident. There were no gender differences in the above perceptions. Most girls and boys had similar views about science and scientists. Both girls and boys saw science as a male activity and scientists as males but only girls drew female scientists (very few) The overall image of science and scientists reflects the existing situation, not the changing scenario. That some girl students manage to break free from the stereotype suggests that even a few inputs to the educational system may help in bringing the desired changes.

Science and technology are intertwined with gender mediated by symbols and language. Several researchers (Gurer and Camp 1998) working in the area of gender, technology and language have pointed out that the language use in technology is 'gendered'. Gender and experiences in school and at home (MacKenzie and Wajcman 1999) as well as interactions with technological artefacts influence attitudes of individuals towards technology. There have been numerous PATT (Pupil's Attitude Towards Technology) studies aimed at finding students' attitudes towards technology (Bame, Dugger, de Vries and McBee 1993).

In 2006, a survey aimed to elicit students' understanding of technology carried out with over 200 students of Class 6 (11–14 year-olds) in schools in and around Mumbai, two questionnaires on 'technology-as-objects' and 'technology-as-activities', were administered followed by interviews of a sub-group (Mehrotra, Khunyakari, Natarajan and Chunawala 2007). The results indicated that objects and activities related to communication and transport, especially modern gadgets used in urban areas were often considered technological. Objects presented along with humans were perceived to be more related to technology than objects by themselves. In interviews students focussed on the usefulness and the human-made aspect of artefacts and acknowledged their role in easing and speeding work. Most interviewed students consistently believed that technology has existed in the recent past, has evolved and is ubiquitous now. Scientists and researchers were credited with creation of technology while others were mere users.

Pedagogical Practices: Gendering the Textbook and the Classroom

The under-representation of women in science is often 'explained' by suggesting that there are biological differences in cognitive ability between men and women. The issue of sex differences in learning falls into the classic argument of nature versus nurture; research in this area has been inconclusive as the differences in ability, if any, appear only at ages when it is difficult to separate the effects of genetic factors from socialization. Thus there may or may not be biological explanations for sex differences in learning but research has highlighted the role of sociological factors, such as differing expectations, differences in learning and attitudes of school boys and girls. Different aspects of the educational system play an important role in building gender identities.

Textbooks, used extensively by teachers and students, play a large role in formal education. In India, there is a great dependence on textbooks mainly because of a lack of other educational materials. A study of textbooks by Narendra Nath Kalia in 1979–80, which analysed 41 books (21 English and 20 Hindi) prepared and used by the National Council of Education Research and Training (NCERT) and Central Board of Secondary Education (CBSE) indicated widespread and extensive gender bias in the textbooks. Not only were women portrayed in very few of the lessons as compared to men, (the ratio being 1:3) whenever women were portrayed, they were depicted as inferior to men. Of the large number of occupations depicted in the textbooks, women were excluded from the majority. The few occupations held by women were generally lower in income and prestige as compared to men (Kalia 1986).

Regarding the use of masculine pronouns like 'he' 'his' or words like 'mankind' 'man', it is often argued that these are merely semantics and children understand that these words refer to both men and women. However, studies have shown that young children, given information of generic language such as 'mankind' and 'he', draw pictures of men and boys when asked to visually present the information or story they had heard (Martyna 1978, cited in Rosser 1992). Although adult women have learned that generic language is inclusive, some studies have shown (Thorne 1979, cited in Rosser 1992) that women feel excluded by such language.

Gender-based subject choices of boys and girls may also be influenced by the portrayal of gender in school science texts. An analysis of illustrations and texts in Indian science textbooks of grades 3-10, and a survey of students' and teachers' ideas related to gender, revealed gender biases by omission and commission (Chunawala, Vinisha and Patel 2009). There were significantly fewer female figures and often these were stereotypical images depicted in non-remunerative occupations limited to the domestic space. Women were rarely shown as contributing to historical or present-day events related to S&T. This evident lack of female role models in the textbooks sets a poor example for young girls who may aspire to be scientists and may discourage the pursuit of science. Similarly, such stereotypical images of men may prevent boys to pursue their dream of becoming elementary school teachers. Further, the study revealed gender stereotyped perceptions regarding occupations among students and teachers as well as the usage of biased terminology like 'mankind', 'man-made', and 'businessmen', and so on, in the textbooks.

Research in relation to classroom interactions of teachers with students has increasingly shown that male students receive more of every type of classroom interaction, that is, they receive significantly more praise and criticism than females (Jones 1989). Boys in school are asked more open-ended questions than girls and they are often given directions on how to do things themselves. On the other hand, teachers often do things for the girls in classrooms. Research has also shown that boys are more assertive in the classroom and more likely to shout out the answers, as a result teachers respond to them more.

Teacher gender has been found to be unrelated to the differences in classroom interaction; both male and female teachers interact in similar ways with their students. An important point to note is that teachers are not consciously discriminating between the students. They are convinced that they are being gender-neutral, despite the fact that observers notice the differential treatment. What this means is that teachers are not intentionally stereotyping students, but their behaviour reflects that they themselves are members of society and products of a biased educational system. Teachers reflect the values and expectations of society. But teacher expectations may affect student achievement as demonstrated by the self-fulfilling prophecy. If teachers can be made to recognize biases they can make many positive changes to the classroom situation that can impact S&T education.

Science, Technology and Education in India

In earlier times, the aim of education was developing the skills needed for making a living, for livelihoods. It was technology-based and involved learning crafts. Apprenticeship was one form of learning and 'vocational' learning too found a place in formal education. However, an elitist and theory-based stream of education also existed, as in the brahmanical tradition (or in Plato's Academy and later church-based education in the West). The tension between mental (science) and manual (technology) education has long existed and this dichotomy reflects what Keller (1985) indicates is the human tendency to think in hierarchical binary categories. This is so in the case of male and female genders, as well as other associated binaries, such as science/nature, mind/body, rational/emotional, objective/subjective–the first term in each binary category being associated with the male and the second with the female in a gendered hierarchy.

Catering to over a million school going children nation-wide, India's education system is torn by several conflicting interests. Besides being low the 'educational achievements are highly uneven' with literacy rates varying by 'region class, caste and gender' (PROBE Team 1999: 11). The educational system ignores the socio-cultural diversities among Indian students and teachers and there is a need to promote a plurality of strategies to address these diversities. The problems of mismatch between culture, identity, educational content and pedagogy are compounded in the teaching of S&T without connection to local contexts.

It is generally accepted in India that science, technology, and education are critical ingredients for national, economic and social development. The growing importance of technology in all spheres of life has made it imperative that we have a formal programme of study for children from a young age. The New Policy on Education–NPE, 1986 (Department of Education 1992) recognized the importance of technology for personal and social development, and the National Curriculum Framework 2000 (NCF 2000) introduced 'Science and Technology' textbooks at the secondary school level. These, unfortunately, presented technology in the limited paradigm of applied science (Ramadas 2003). Technology is a domain in its own right, different from science, and yet uses several forms of knowledge including scientific knowledge (Staudenmaier 1989)

Broadly defined, technology has the potential to be a component of existing school subjects (NCF 2000): subjects like 'Art of Healthy and Productive Living' at the primary classes, and 'Work Education' and 'Art Education' at the upper primary. The 'Work Education' subject for higher classes translates as pre-vocational courses for knowledge and skills to enter the world of work. However, the importance accorded to them in the scheme of school education makes these subjects irrelevant.

The National Curricular Framework 2005 (NCF 2005), for the first time referred to design and technology as part of teaching and learning science at the school level. Its Position Paper of the National Focus Group on Teaching of Science (NFG-ToS 2006: 2), for the first time in the history of curricular frameworks, recognized that

> Technology as a discipline has its own autonomy and should not be regarded as a mere extension of science. . . . Technological solutions are guided as much by design, aesthetic, economic and other practical considerations as by scientific principles.

Besides, the Position Paper of the National Focus Group on Work and Education, 2007 (NFG-W and E 2007: 30) linked the role of productive work and design opportunities in education: 'A systematic study of design and technology can provide opportunities for learning a broad spectrum of generic skills and competences.'

There are several facets to technology education at the school level. As an economic instrument, technology education contributes to national economic competitiveness and wealth creation and, in that restricted sense, is synonymous with vocational education. Technology education may be designed to enhance people's ability to control technology and resist the prospects of the technocratic élite. It has the potential to either counter gender biases in the present-day representation of technology or, if inappropriately handled, to perpetuate them. According to Murchland (1982: 5),

> If one regards technology as a language (and not merely as a tool) and technicism as a perversion of that language, then the function of education becomes clear. For the primary task of education is to train us in the responsible use of language . . . Put simply, education is what enables us to create and control the symbolic worlds, including, of course, the world of technology. That is what literacy means.

A carefully designed curriculum for design and technology will enable students with all kinds of abilities and backgrounds to participate in sustainable living. The planning of the design and technology curriculum includes making several choices: the nature of technology to be addressed, who it will be designed for and by whom, the structuring of classroom processes of design (ideation/conceptualization) and making (actualization). The materials for the tasks can be planned to be chosen in the context of the task and by the participants influenced by their own knowledge and values. The sensitivities of resource use as well as social and material sustainability can be integrated into all decision making.

According to Dias (2004), diverse and complex forces have shaped present day education–especially with regard to what is considered as valid and valued. The obsession with the 3 Rs (reading, writing and arithmetic) has pushed several areas of human skills and knowledge out of education. There is no place within education for, or even tolerance of, different forms of human expressions such as drawing, enactment, making and other skill-based expressions.

As it has evolved, technology education provides opportunity for students to learn about the processes and knowledge needed to solve problems and extend human capabilities. Of the various levels at which technology education is introduced into education, in India higher education has been the most preferred level in terms of resource allocation, as in the Indian Institutes of Technology (IITs) and the Industrial Technical Institutes (ITIs). Yet, as already mentioned there are very few women at this level (graduation and post-graduation) in engineering and technology (Parikh and Sukhatme 2004).

Along with other socio-cultural reasons that prevent women's employment in technological occupations, the introduction of technology at the tertiary stage rather than in school could be an important reason for the skewed participation of women in technology. By the tertiary level, the gender stereotyping of professions and stereotypical distribution of students at intake are already in place. The appropriate place to challenge the existing practices of technology, including gender aspects, is not at the higher education levels but at school. Introduction of technology at school needs to be considered in detail as it would be accompanied by its own set of complexities.

Introducing Technology Education in the General School Curriculum

Technology education at the school level in India has emerged in several forms like vocational education and socially useful and productive work (SUPW). Besides, it has been stereotyped on the basis of gender (Chunawala 2004) with topics such as needlework for girls and bookbinding for boys.

According to Layton (1993) many stakeholders have played an important role in placing technology education within the general school curriculum. Mehrotra (2008) points out that technology education when present in school curricula tends to emphasize the male perspective. Welty (1996) reports that the content in technology courses pays attention to the majority who enrol in these courses, namely, boys who value abstraction and competition. Other studies have reported that girls value technology for its role in facilitating collaboration, communication, and linkages between people (Gilligan 1982; Honey et al. 1991). Feminists and other educators have argued that sex bias in education lowers girls' self-esteem and steers them to traditional female courses and careers, but also burdens boys (AAUW 1992). Various ways have been suggested to overcome the problem of alienation of girls, such as restructuring of subject matter, revising language by paying closer attention to explanation and context, creating humane classroom environments and valuing a variety of ways of knowing, expressing and working, integrating cognitive and affective learning and discussing values related to technology (Zuga 1999).

Technology education has not only failed girls, it has also failed other marginalized groups: students from rural areas, tribal communities, the poor and those who drop out of school due to a variety of reasons. One of the reasons is the alienation of school knowledge from everyday life, which happens at various levels. Most members of marginalized groups are creative in designing a sustenance for themselves and their families from the limited resources available to them. Technology education that begins only at the tertiary stages leaves out of its ambit a large group of students disenchanted with the process of getting an education that does not address everyday problem-solving and their own sustenance. Besides, technology, as defined hierarchically from vocational to engineering education, fails to even recognize the technologies created by the marginalized. Introduction of technology at school has the potential to meet the concerns of equity in access to technological knowledge, processes and activities, where all would be introduced to technology and the everyday technology of all would gain credibility in the process of schooling. For this to happen, it is also imperative that technology education be introduced early in school, be inclusive and collaborative, and allow different forms of communication.

Attempts in the direction of making technology education inclusive have been made by our research group through the development of collaboration and communication-centred design and technology units for middle school students. This aimed at providing inputs to plan technology education curriculum for Indian students and helping teachers and planners to know the ideas that children hold of technology. A prior survey had suggested that Indian students' ideas of technology, though varied, lack depth (Mehrotra et al. 2007). Their view of technology is rooted in science, either as its applications or as its object of study.

There is a need to introduce technology at school as a subject with distinct knowledge and skill requirements. Teachers and educators need to know the multiple perspectives of technology so that in their classrooms they may be able to make appropriate linkages of technology with science and society as well as with other school subjects.

Towards an Inclusive Technology Education: Communication, Collaboration and Creativity

Technology education has to be introduced in schools and it has to be completely restructured so as to make it inclusive. The 'add women/marginalized groups and stir' methods are not successful. The curriculum needs to focus on people and values associated with technology. Collaboration and communication are styles of working that research shows are preferred by women and these need to be integrated into technology education.

The social interactions that occur in group work require transmission of ideas, thoughts and emotions to one another through verbal and non-verbal ways. Vygotsky (1978) emphasized social context and the role of language in his theory of development of cognitive functioning. According to him, jointly undertaken, goal-oriented activities are important for learning, and language is a major psychological and cultural tool for representing ideas, interpreting and evaluating events and experiences, and constructing explanations. The ability to use language is central to an individual's overall development and especially in developing technological capability (Rowell 2002).

Collaborative learning is based on the premise that learning is best achieved in interactions rather than through individual or one-way transmission process (Haller et al. 2000). Studies have reported that collaborative learning involves communication, discussion, negotiation and construction of new knowledge (Baker 1999; Medway 1994). The attempt at questioning, answering, elaborating, explaining and verbalizing new ideas, helps learners express their thoughts (Chi et al. 1989; Wegerif and Mercer 1996; Weiss and Dillenbourg 1999).

Despite the demonstrated importance of communication, socialization and teamwork for all-round development, there is limited appreciation that skills needed for collaboration need to be deliberately fostered in the context of classroom activities. Design and technology activities provide a potentially rich environment for fostering collaborative learning–both for expression and accommodation of individual perspectives as well as opportunities for group work. In design and technology, the ideas conceived in the mind need to be expressed in concrete form before they can be examined to see how useful they are (Kimbell et al. 1991). Researchers have pointed to the need to link technology with its social implications as the enterprise of technology involves various groups of people–the clients, designers, makers and users–who form a 'community of practice' (Wenger 1998). Community of practice refers to a group of people that share a common interest in a domain of knowledge and skills and the term has gained acceptance as participation in communities of practice form a basis for learning. As a broader concept, the members of such communities have an opportunity to develop themselves personally and professionally through the process of sharing information and experiences with the group. Besides, the members experience and create their shared identity through engaging in and contributing to the practices of their communities.

Cognitive activity is tied to the social context in which it occurs (Natarajan 2007) and therefore classroom activities need to be contextualized in order to make them inclusive. Research has shown that students from rural and urban areas and girls and boys have different learning styles: rural students tend to be more 'serious analytical learners and active practical learners' as compared to urban students (Cox, Sproles and Sproles 1988) and girls prefer collaboration over competition (Honey 1994). Contextualizing activities provides all groups of students, irrespective of their social and educational setting, opportunities to employ an empowering range of knowledge, skills and values. Technology education activities also offer opportunities to visualize and creatively redesign the environment in ways that can be meaningful to all. By its very nature technology is diverse and provides possibilities for students to engage in a wide variety of tasks depending on their choice and aptitudes and is ideal to reach out to girls and marginalized students.

Close to 200 million children in India are in the school-going age group of between five and 18. A very large number of these children have either never enrolled in schools or have dropped out at various stages. Yet, those who drop out of the education system join (or aspire for) the world of work, most without acquiring employable skills. What does school education provide in terms of employable skills like knowledge, process, procedures or team work? With an emphasis on studying theoretical principles and observing experiments conducted primarily by the teacher, school science is seldom experienced as an activity at all, leave alone a participatory one. A study suggested that Indian students do not view science as a collaborative activity (Chunawala and Ladage 1998). Work experience, in any of its variations at school, generally involves making socially useful objects using recipes suitable for production in large numbers. It has little scope for design, creativity or for examining contexts of use.

Technology education can involve multidisciplinary perspectives and multiple skills. If school education emphasizes mere technical literacy it stifles technological innovations. Enriching school curricula with explicit opportunities for authentic problem-solving and multiple expression modes valid in a variety of classroom contexts could help develop future technology innovators in a diverse population. Technology education is more than just vocational education or work experience. It is discipline that is grounded in science and is linked to arts, crafts, economics and traditional knowledge systems too (Natarajan and Chunawala 2009). According to Cross (2002), design is at the core of any technological activity, and views the world differently from science and humanities. Hence technology, with an emphasis on design, as a subject in Indian general education is likely to appeal to a wider cross-section of students than do either the sciences or humanities and arts. Thus, technology, in general education, by being accessible to all and tapping a broader base of potential innovators, can uphold the two aims of equity and excellence in education.

Summary of Current Scenario and Future Directions

In India, the participation of women in engineering has been negligible. According to Parikh and Sukhatme (2004) women comprised about 15 per cent of the total graduates from major engineering institutes and of those who graduate, over 30 per cent remain unemployed. And these numbers would often be from urban, middle class families. The vocational education and training programmes of the country have never adequately met the needs of the rural and marginalized groups, of which women are a part. On the one hand, innovations through the national research and development (R&D) spending have not touched the lives of the marginalized, whose innovations for sustenance have failed to attract funding and guidance. The reasons for this state of affairs are complex and arise from academic, organizational, economic and socio-cultural contexts.

The National Knowledge Commission (NKC), constituted by the Government of India in 2005 to study the various aspects of education, at different levels, has recommended more flexibility in vocational education and training (VET). It has proposed a complete overhaul of the regulatory structure for higher education, with the creation of an Independent Regulatory Authority for Higher Education. For technical and vocational education to gain status, it will need teachers with better professional preparation. This implies teaching elements of design and critical thinking about technology and society not only to those in the vocational education stream, but also to all students in the generic vocational courses.

Making technology education inclusive assumes significance in the complex socio-cultural context of India: immense innovation potential across the country among the schooled and the unschooled, the formal and non-formal sectors; continuing traditions of indigenous local technologies; wide cultural and resource differences between different parts of the country including the rural-urban divide, and that between the female population and their participation in S&T in classrooms and in society. This is extremely important considering, that of the students enrolling in Class 1, a large proportion drop out by Class 8 and are involved in the world of work. Thus to have a workforce that is both creative and productive too, there is a need to introduce a contextualized an equitable design and technology (D&T) education from the primary years of schooling.

To reiterate Natarajan and Chunawala (2009: 115),

> The present education system that does not encourage collaboration and constrains modes of expression can be alienating to a large proportion of learners. The D&T model developed at the Homi Bhabha Centre for Science Education (HBCSE) proposes stage-wise increasing complexity of activities, which integrate skills and processes across knowledge domains. A well-planned D&T curriculum in mixed ability and multicultural classrooms can be an inclusive rather than an exclusive endeavour—for the children of the rich as well as the dispossessed, for those in the indigenous or the modern mould, for girls and for boys.

The educational community tends to be conservative and does not easily adopt radical changes. Encouraging creativity and critical thinking and assessing them are not simple tasks in classroom situations. Bringing these changes can threaten the status quo of the existing social systems (ibid.). Also technology education is a broader and more ambiguous discipline in contrast to engineering education.

The multiple manifestations of technology—as knowledge, design and making in the context of human needs and desires—facilitates intimate connections between thought and action within the socio-cultural context in which it arises (Mitcham 1994). What is technology varies with time and context, and this feature of the discipline provides scope for inclusivity, equity, creativity and critical thinking. We urge the inclusion of design and technology in Indian school curricula and hope that all the empowering possibilities are realized some day.

References

AAUW. 1992. *How Schools Shortchange Girls*. Report of the American Association of University Women Education Foundation. Washington DC: AAUW.

Baker, M. 1999. 'Argumentation and Constructive Interaction'. In *Foundations of Argumentative Text Processing*, edited by J. Andriessen and P. Coirier. Amsterdam: University of Amsterdam Press: 179–202.

Bame, E., W. Dugger, M. de Vries and J. McBee. 1993. 'Pupils' Attitudes Toward Technology-PATT-USA', *Journal of Epsilon Pi Tau* 12, 1: 40–48.

Chambers, D. W. 1983. 'Stereotypic Images of the Scientist: The Draw-a-Scientist-Test', *Science Education* 67: 255–65.

Chi, M. T. H., M. Bassok, M.W. Lewis, and P. Reimann et al. 1989. 'Self-Explanations: How Students Study and Use Examples in Learning to Solve Problems', *Cognitive Science* 13: 145–82.

Chunawala, S. 2004. 'Education and Technology Education within the Gender Perspective'. In *Multiple Languages, Literacies and Technologies*, edited by P. V. Dias. New Delhi: Books for Change: 162–77.

Chunawala, S., and S. Ladage. 1998. *Students' Ideas about Science and Scientists*, Technical Report No. 38, Mumbai: Homi Bhabha Centre for Science Education.

Chunawala S., K. Vinisha and A. Patel. 2009. *Gender, Science and Schooling: Illustrations in Science Textbooks and Students' and Teachers' Ideas Related to Gender*. Mumbai: Homi Bhabha Centre for Science Education.

Cockburn, C. 1985. *Machinery of Dominance: Women, Men and Technical Know-How*. London: Pluto.

———.1983. *Brothers: Male Dominance and Technological Change*. London: Pluto.

Cox, D. E., E.K. Sproles and G.B. Sproles. 1988. 'Learning Style Variations between Rural and Urban Students', *Research in Rural Education* 5, 1: 27–31.

Cross, N. 2002. 'The Nature and Nurture of Design Ability'. In *Teaching Design and Technology in Secondary Schools*, edited by G. Owen-Jackson. London: Routledge: 124–39.

Department of Education. 1992. *National Policy on Education 1986: Revised NPE and Programme of Action*. New Delhi: Ministry of Human Resource Development, Government of India.

Dias, P.V. 2004. 'Diversity of Knowledge Production through Multilingual Relationships, Multiple Literacies and Appropriate Technologies'. In *Multiple Languages, Literacies and Technologies*, edited by P. V. Dias. New Delhi: Books for Change: 1–12.

Faulkner, W. 2001. 'The Technology Question in Feminism: A View from Feminist Technology Studies', *Women's Studies International Forum* 24, 1: 79–95.

Gilligan, C. 1982. *In a Different Voice: Psychological Theory and Women's Development*. Cambridge, MA: Harvard University Press.

Gurer, D., and T. Camp. 1998. 'Investigating the Incredible Shrinking Pipeline for Women in Computer Science (Final Report)' National Science Foundation Project 9812016.

Haggerty. S. M. 1995. 'Gender and Teacher Development: Issues of Power and Culture', *International Journal of Science Education* 17, 1: 1–15.

Haller, C. R., V. J. Gallagher, T.L. Weldon, and R.M. Felder. 2000. 'Dynamics of Peer Education in Cooperative Learning Work Groups'. *Journal of Engineering Education* 89, 3: 285–293.

Harris, M. 1997. *Common Threads: Women, Mathematics and Work*. Stoke-on-Trent, UK: Trentham Books.

Honey, M. 1994. 'The Maternal Voice in the Technological Universe'. In *Representations of Motherhood*, edited by D. Bassin, M. Honey and M. M. Kaplan. New Haven, CT: Yale University Press: 220–39.

Honey, M., B. Moeller, C. Brunner et al. 1991. *Girls and Design: Exploring the Question of Technological Imagination*. New York: Center for Technology in Education, Bank Street College of Education.

INSA. 2004. 'Careers for Indian Women: An Examination of Indian Women's Access to and Retention in Scientific Careers.' Report. New Delhi: Indian National Science Academy.

Jones, G.M. 1989. 'Gender Bias in Classroom Interactions', *Contemporary Education* 60, 4: 218–22.

Jones, G. M. and J. Wheatley. 1988. 'Factors Influencing the Entry of Women into Science and Related Fields', *Science Education* 72, 2: 127–42.

Kalia, N. 1986. 'Women and Sexism: Language of Indian School Textbooks', *Economic and Political Weekly* 21, 18: 794–97.

Keller, E. Fox. 1985. *Reflections on Gender and Science*, New Haven, CT: Yale University Press.

Kimbell, R., K. Stables, T. Wheeler et al. 1991. *The Assessment of Performance in Design and Technology.* London: School Examinations and Assessment Council.

Layton, D. 1993. *Technology's Challenge to Science Education.* Milton Keynes: Open University Press.

MacKenzie, D., and J. Wajcman, eds. 1999. 2nd ed. *The Social Shaping of Technology.* Milton Keynes: Open University Press.

Martyna, W. 1978. 'What Does He Mean? Use of the Generic Masculine', *Journal of Communication* 28: 131–38.

Mead, M., and R. Metraux. 1957. 'Image of the Scientist among High School Students', *Science* 126: 384–90.

Medway, P. 1994. 'The Language Component in Technological Capability: Lessons from Architecture', *International Journal of Technology and Design Education* 4: 85–107.

Mehrotra, S., R. Khunyakari, C. Natarajan and S. Chunawala. 2007. 'Using Pictures and Interviews to Elicit Indian Students' Understanding of Technology'. In *Teaching and Learning Technological Literacy in the Classroom: Pupils' Attitudes Towards Technology*, edited by J. R. Dakers, W. J. Dow and M. J. de Vries *(PATT 18) Conference.* June 2007, Glasgow: Faculty of Education, University of Glasgow: 152–61.

Mehrotra, S. 2008. 'Introducing Indian Middle School Students to Collaboration and Communication Centred Design and Technology Education: A Focus on Socio-Cultural and Gender Aspects'. PhD Dissertation, HBCSE, Mumbai.

Mitcham, C. 1994. *Thinking through Technology: The Path between Engineering and Philosophy*. Chicago: University of Chicago Press.

Murchland, B. 1982. 'Technology, Liberal Learning and Civic Purpose', *Liberal Education* 68, 4; cited by W.B. Waetjen. 1987. 'The Autonomy of Technology as a Challenge to Education', *Bulletin of Science, Technology and Society* 7, 1 & 2: 28–35.

Natarajan, C. 2007. 'Culture and Technology Education'. In *Analyzing Best Practices in Technology Education*, edited by Marc de Vries. Dordrecht: Sense Publishers: 153–67.

Natarajan, C., S. Chunawala, S. Apte and J. Ramadas. 1996. Students' Ideas about Plants, Diagnosing Learning in Primary Science (DLIPS) Part-2, Technical Report No. 30. Mumbai: HBCSE.

Natarajan, C., and S. Chunawala. 2009. 'Technology and Vocational Education in India'. In *International Handbook of Research and Development in Technology Education*, edited by Alistair Jones and Marec de Vries. Dordrecht: Sense Publisher: 105–16.

NCF. 2000. *National Curriculum Framework for School Education*. New Delhi: National Council of Education Research and Training, New Delhi.

NCF. 2005. *National Curriculum Framework*. New Delhi: National Council of Educational Research and Training.

Newton D.P., and L. D. Newton. 1992. 'Young Children's Perceptions of Science and the Scientist', *International Journal of Science Education* 14, 3: 331–48.

NFG-ToS. 2006. 'Paper of the National Focus Group on the Teaching of Science', New Delhi: National Council of Education Research and Training.

NFG-W&E. 2007. 'Position Paper of the National Focus Group on Work and Education'. New Delhi: National Council of Education Research and Training.

NFG-G&E. 2006. 'Position Paper of the National Focus Group on Gender Issues in Education'. New Delhi: National Council of Education Research and Training.

NPE.1986. 'National Policy on Education'. New Delhi: Ministry of Education, Government of India.

Oudshoorn, N., A. R. Saetnan and M. Lie. 2002. 'On Gender and Things. Reflections on an Exhibition on Gendered Artifacts', *Women's Studies International Forum* 25, 4: 471–83.

Parikh, P. P., and S. P. Sukhatme. 2004. 'Women Engineers in India', *Economic and Political Weekly* 39, 2: 193–201.
PROBE Team (Anuradha De, Jean Drèze, A. K. Shiva Kumar et al.). 1999. *The Public Report on Basic Education in India*. New Delhi: Oxford University Press.
Ramadas, J. 2003. 'Science and Technology Education in South Asia'. In *Innovations in Science and Technology Education*, edited by Edgar W. Jenkins, vol. 8. Paris: UNESCO: 95–122.
Rowell, P. M. 2002. 'Peer Interactions in Shared Technological Activity: A Study of Participation', *International Journal of Technology and Design Education* 12: 1–22.
Rosser, S. 1992. *Biology and Feminism: A Dynamic Interaction*. New York: Twayne Publishers.
Sadker, M., D. Sadker and M. Donald. 1989. 'Subtle Sexism at School', *Contemporary Education* 60, 4: 204–12.
Schiebinger, L. 1989. *The Mind Has No Sex? Women in the Origins of Modern Science*. Cambridge, MA: Harvard University Press.
Sjoberg, S., and G. Imsen. 1988. *Gender and Science Education: In Development and Dilemmas in Science Education*, edited by P. Fensham, London: Falmer: 218–48.
Solomon, J., J. Duveen and L. Scott. 1994. 'Pupil's Images of Scientific Epistemology', *International Journal of Science Education* 16, 3: 361–73.
Staudenmaier, J. M. 1989. *Technology's Storytellers: Reweaving the Human Fabric*. Cambridge, MA: MIT Press.
Thorne, B. 1979. 'Claiming Verbal Space: Women Speech and Language in College Classrooms.' Paper presented at the Research Conference on Educational Environments and the Undergraduate Woman, Wellesley College, Wellesley, MA: September.
Vygotsky, L. S. 1978. *Mind in Society: The Development of Higher Psychological Processes*. Cambridge: MA: Harvard University Press.
Wajcman, J. 1991. *Feminism Confronts Technology*. Oxford: Blackwell.
Weber, K., and R.L. Custer. 2005. 'Gender-Based Preferences toward Technology Education Content, Activities, and Instructional Methods', *Journal of Technology Education* 16, 2: 55–71.
Wegerif, R., and N. Mercer. 1996. 'Computers and Reasoning through Talk in the Classroom', *Language and Education* 10, 1: 47–64s.

Weiss, G., and P. Dillenbourg. 1999. 'What Is "Multi" in Multi-Agent Learning?' In *Collaborative Learning, Cognitive and Computational Approaches*, edited by Pierre Dillenbourg, Oxford: Elsevier: 1–20.

Welty, K. 1996. 'Identifying Women's Perspectives on Technology'. Paper presented at the 58th Annual Conference of the International Technology Education Association, 31 March–2 April, Phoenix, Arizona.

Wenger, E. 1998. 'Communities of Practice. Learning as a Social System', *The Systems Thinker* 9, 5. https://thesystemsthinker.com/communities-of-practice-learning-as-a-social-system/ accessed 28 December 2016.

Winner, L. 1986. *The Whale and the Reactor: A Search for Limits in an Age of High Technology*. Chicago: University of Chicago Press.

Zuga, K. 1999. 'Addressing Women's Ways of Knowing to Improve the Technology Education Environment for All Students', *Journal of Technology Education* 10, 2: 57–71.

About the Editors and Contributors

Sumi Krishna is a distinguished independent scholar and former President of the Indian Association for Women's Studies. She has over 40 years of experience in environment, development and gender, encompassing biodiversity, natural resource management, people's movements and livelihood issues; has advised universities and institutions on integrating science and social science curricula and methodologies. She is a widely published author and is based in Bengaluru.

Gita Chadha is Assistant Professor, Department of Sociology, University of Mumbai. She has taught for over 10 years at the undergraduate level; designed and taught the first Feminist Science Studies course at TISS, Mumbai; designed pedagogic initiatives for integrating science and social science teaching conducted by the Centre for the Study of Culture and Society, Bengaluru.

Contributors

Asha Achuthan is Assistant Professor, Advanced Centre for Women's Studies, Tata Institute of Social Sciences, Mumbai. After qualifying as a medical doctor, she took up questions of development and Third World women and has a PhD on feminist critiques of science, focusing on India. She analyses the histories and self-descriptions of the sciences and the ways in which the practice of science, particularly the discourse of Community Medicine, has changed.

Anitha Kurup is Professor, School of Social Sciences, National Institute of Advanced Studies, Bangalore, and anchors the education programme, including the National Programme on Gifted Education. In 2011–12, under a Fulbright Nehru Fellowship, at the University of California, Davis, she carried out a comparative study of women scientists and engineers in India and the USA. In 2014, she was invited to conduct a gender workshop at the International Conference of Women in Physics, Waterloo, and was a co-author of the country paper.

Chayanika Shah is a feminist and queer rights activist. She has campaigned, researched, taught and written on the politics of population control and reproductive technologies; feminist studies of science; communalism; sexuality and sexual rights. Her co-authored books include a companion volume for the documentary *Bharat ki Chaap*; and *We and Our Fertility: The Politics of Technological Intervention*. She has a doctorate in physics and has taught at a Mumbai college for over 20 years. She has also designed and co-taught courses on 'Feminist Science Studies' and on 'Science Education' at the Tata Institute of Social Sciences, Mumbai.

Chitra Natarajan (late) was Professor and Dean, Science Education Faculty, Homi Bhabha Centre for Science Education (HBCSE), TIFR, Mumbai. She was an active researcher in design and technology education and environmental education, and her interests included socio-scientific issues, project-based learning and students' and teachers' reasoning about design and systems. She edited a series of eight books on science, technology and society, and there are over 80 publications to her credit. She was a member of the editorial boards of *International Journal of Technology and Design Education*, the Netherlands, and *Design and Technology Education: An International Journal*, UK.

Ipshita Chanda is Professor, Department of Comparative Literature, and Professor, Centre for African Literatures and Cultures, Jadavpur University, Kolkata. Her publications in Gender Studies are: *Shaping the Discourse: Women's Writings in Bengali Periodicals, 1864–1947* (2013), *Selfing the City: Women Migrants and Their Lives* (forthcoming) and the entry on 'Third World Feminism' in Ray and Schwarz, eds, *Wiley Encyclopedia of Postcolonial Studies* (forthcoming). She has worked on several translations from Hindi and Bengali into English.

Kamala Ganesh has retired as Professor, Department of Sociology, University of Mumbai. She has written on gender and kinship, feminist archiving, culture and identity and the Indian diaspora. Among her several publications, *Boundary Walls: Caste and Women in a Tamil Community* won the silver medal of the Asiatic Society of Bombay; and her jointly edited volume *Zero Point Bombay: In and around Horniman Circle* was listed by the *Guardian* as among the ten best books set in Mumbai.

Kanchana Mahadevan is Associate Professor, Department of Philosophy, University of Mumbai, specializing in feminist philosophy, continental philosophy, epistemology, ethics and socio-political philosophy, including analytic philosophy, environmental philosophy and ancient Greek philosophy. She is also interested in the interdisciplinary fields of diaspora and film studies. Under the aegis of the Indian Council of Philosophical Research, she has prepared 'Between Femininity and Feminism: Philosophical Perspectives' and is involved in introducing feminism in the philosophy curriculum.

Madhumita Mazumdar is Associate Professor at Dhirubhai Ambani Institute of Information Communication Technology, Gandhinagar. She has a specialized interest in the social and cultural history of science and technology in colonial and contemporary India. She has been involved in an ESRC (UK) funded collaborative research project titled, 'Integrated Histories of the Andaman Islands' where she explores the historical legacies of colonial science in the shaping of contemporary developmentalist agendas and tribal welfare policies. Much of her published work derives from her early researches into science and the cultural politics of nationalism in colonial Bengal.

Meena Gopal is Associate Professor and former Head, Department of Women's Studies, Tata Institute of Social Sciences, Mumbai. Formerly, she was with the Research Centre for Women's Studies at the SNDT University, Mumbai. She is also an activist with the Bombay-based feminist collective, Forum against Oppression of Women.

About the Editors and Contributors

Meghana Kelkar, an officer in the Department of Agriculture, Maharashtra, is based in Pune. In 1966, she was the first woman to be directly inducted as a 'Class I Gazetted Officer'. In 2002, she was associated with the M.S. Swaminathan Committee to draft an Action Plan for agricultural development in Maharashtra for the following 25 years. In 2008, she obtained her PhD from the Tata Institute of Social Sciences, Mumbai.

Sugra Chunawala is Associate Professor, Homi Bhabha Centre for Science Education (HBCSE), Tata Institute of Fundamental Research, Mumbai. Her research and extensive publications focus on attitudes of youth towards science, socio-cultural aspects of education and gender in relation to S&T education. She was a member of the Gender Focus Group in the 2005 National Curriculum Framework and is responsible for a permanent exhibition at HBCSE on Gender and Science. Currently, she is involved in research on Design and Technology education and socio-scientific issues. She coordinates a postgraduate distance learning research programme and is currently conducting undergraduate and teachers' orientation courses.

Unnati Tripathi has a Masters in Sociology from the University of Mumbai. She has worked as translator, researcher and co-investigator for *Behind the Beautiful Forevers* by journalist Katherine Boo. She has also assisted on a project to establish the archive of the Indian Association for Women's Studies.

Table of Contents, Vol I

Preface

Introduction

    Understanding Gender and Science in India: Institutions and Beyond
    Sumi Krishna

    Tracking a Consciousness: Questions, Dilemmas and Conundrums of Science Criticism in India
    Gita Chadha

1. *Feminists Discuss Caste and Gender in Science: An Online Dialogue*
   Anita Mehta, Chayanika Shah, Gita Chadha, Mary E. John, Mina Swaminathan, Prajval Shastri and Sumi Krishna

2. *Unravelling the 'Gender-Merit' Conundrum: Do Women Deserve to Do Science in India?*
   Jayasree Subramanian

3. *Re-Cognizing Gender Bias in the Life Sciences*
   Sumi Krishna

4. *The Science of Psychology: Where Is Gender?*
   U. Vindhya

5. *Science, Gender and Reproductive Technologies: A Case of Disability*
   Anita Ghai and Rachana Johri

6. *Gender Inequities in the Science Workplace: An Experiential Perspective*
   Prajval Shastri

7   *A Gender-Sensitive Practice of Psychiatry in India? The Story of Ajita Chakraborty*
Mandira Sen

8   *Women Water Professionals in the Maharashtra Water Bureaucracy*
Seema Kulkarni

9   *Women, Livestock and Rural Livelihoods: Challenges for Veterinary Scientists*
Sagari R. Ramdas

10   *Bridging the Gap between Natural Sciences and Gender Studies: Notes on a Pedagogical Experiment*
Mina Swaminathan

11   *Integrating Gender into the Curricula of Health Professionals: Experiences and Reflections*
T. K. Sundari Ravindran

12   *Teaching Feminist Science Studies in India: An Experiment*
Chayanika Shah and Gita Chadha

13   *En-Gendering Bodies of Knowledge: Scientific Institutions and the Production of Science in Science Fiction*
Suchitra Mathur

List of Contributors
Table of Contents, vol 2